THE FARMER AND THE

This sounds like the title of a modern fable. While inter-changing ideas the farmer and the obstetrician realise to what extent they both manipulate the laws of nature. They analyse the striking similarities between *the industrialisation of farming and the industrialisation of childbirth*, which developed side by side during the twentieth century.

In both cases an innovation was usually presented as the long-awaited solution to an old problem. For example the advent of powerful synthetic insecticides has overnight dramatically reduced costs and increased agricultural pro-ductivity. One can easily understand the reasons for their routine widespread use. Similarly, the advent of the modern safe technique of caesarean section offered serious new reasons to create gigantic obstetrical departments, so that all women could give birth close to operating rooms and specialised medical teams. But soon after the immediate enthusiastic reactions, a small number of sceptics expressed doubts and voiced fears concerning probable negative long-term consequences of the widespread use of novel little-tested attitudes or practices. Although repeated warnings went apparently unheeded, they motivated the development of alternative attitudes. Moreover they became the roots of organised movements involving increasing numbers of con-sumers.

At the turn of the century the history of industrialised farming suddenly speeded up. A collective global awareness was sparked by a series of disasters, particularly mad cow and foot-and-mouth diseases. In contrast, industrialised childbirth has not yet reached the same phase of its history. *Which disaster are we waiting for?*

Also by Michel Odent
and published by Free Association Books

THE SCIENTIFICATION OF LOVE 2nd edition (2001)

Michel Odent is well known as the obstetrician who introduced, in a French state hospital, the concepts of home-like birthing rooms and birthing pools. He is the founder of the Primal Health Research Centre in London, whose objective is to study correlations between what happens during the primal period (from conception until first birthday) and health and behaviour later on in life. He is the author of about 50 scientific papers and 10 books published in 19 languages.

THE FARMER AND
THE OBSTETRICIAN

MICHEL ODENT

'an association in which the free development of each
is the condition of the free development of all'

FREE ASSOCIATION BOOKS / LONDON / NEW YORK

First published 2002 by
Free Association Books Limited
57 Warren Street, London W1T 5NR

www.fa-b.com

© Michel Odent 2002

The right of Michel Odent to be identified as the author
of this work has been asserted by him in accordance with the
Copyright, Designs and Patents Act 1988.

ISBN 978-1-85343-204-0

A catalogue record for this book is available from the
British Library.

Reprints: 10 9 8 7 6 5 4 3 2 1 0
Designed and produced for Free Association Books by
Chase Publishing Service, Fortescue, Sidmouth EX10 9QG

Printed and bound in Great Britain by
CPI Antony Rowe, Chippenham and Eastbourne

CONTENTS

DEDICATION

This book is dedicated to **Roland Chevriot** (1936–1988), one of the most influential men of the twentieth century ... according to the criteria of the twenty-first century.

In 1977, Roland Chevriot invited me to speak, near Paris, at the annual congress of the ecological group Nature et Progres. My brief was to talk about natural childbirth. The other participants were farmers discussing organic farming. I realised that for Roland these two seemingly different topics were part and parcel of the same issue. We talked about the dangers of playing God, that is going beyond the safety limits of our domination of the laws of nature. The farmers I listened to at this conference were different from those I was accustomed to meeting in the rural area where I practiced. They were endowed with an obvious deep-rooted respect for Mother Earth. Ever since that day one basic question continues to dominate my work: How does the respect for Mother Earth develop?

For Roland and his wife Monica, organic farming was a part of a broad context. It was an aspect of a whole lifestyle. In 1976 they asked me to attend at home the birth of their last child. Normally, I should have turned down their request. At that time, as a full-time practitioner, I was not permitted to practise outside our hospital. However I was allowed to visit them as a friend. This is how it came about that I was witness to the birth of Anne, on 14 June, just after midnight, in the moonlight, while the windows were wide open and the peacocks were screeching in the park. I discovered home birth.

In 1969, Roland was one of the most active members of a small association of eccentric and marginal French amateurs who wanted to promote organic farming. Unfortunately all the other active members were killed in a car crash. This is how Roland, a creative researcher in metal casting, became the soul and the motor of Nature et Progres. In 1972, he went to the USA and met Jerome Rodale, the author of *Pay Dirt*, which is the classic text of the American organic canon. Rodale was also the publisher of *Organic Farming and Gardening*; he cooperated with Ehrenfried Pfeiffer, the leading advocate of biodynamic cultivation after the death of Rudolf Steiner, and was influenced by agricultural thinkers of huge stature such as Robert McCarrison and Sir Albert Howard.

When he returned to France, Roland was thoroughly convinced of the urgent necessity to create a federation of the multitude of small groups involved in organic farming all over the world. Without any delay he pulled the strings to organise a conference. The conference took place in November 1972, at the Palais des Congres, in Versailles. It attracted a core of highly motivated individuals from countries such as France, the USA, the UK, Germany, Sweden and South Africa. This is how the International Federation for Organic Agriculture Movements (IFOAM) was born. Over the years IFOAM became more and more influential in promoting and lobbying the organic cause on an international level. In the 1980s IFOAM was granted official status with the United Nations Department of Information, and cooperation with the Food and Agriculture Organisation (FAO) started. The World Health Organisation (WHO), World Bank, the World Trade Organisation (WTO), Greenpeace, the World Wide Fund for Nature (WWF), Via Campesina and many other international institutions and non-governmental organisations became dialogue partners.

It was also in the 1970s that Roland found it necessary to introduce the concept of the 'ecological fair'. As a visionary and as a doer, he organised in 1976 the first 'Marjolaine' exhibition in Paris. The ecological fair was much more than a market place. It was also a place for discussions and conferences. The room was too small to contain the large number of people who attended our Sunday afternoon conference about childbirth. Roland's objective was consciousness-raising among the general public. Since 1976 there have been Marjolaine exhibitions every year. Today the Paris fair is an important national event and there are similar exhibitions in other French cities.

In the 1980s Roland had another ambitious project. He obtained permission to occupy and use a castle owned by the French administration. In le chateau de Chamarande, near Paris, he wanted to create the 'university of the twenty-first century'. Unfortunately the project was probably premature. It did not attract the support that it would have deserved. One of the last important events that took place in the castle was a conference about human beings and water, in 1983. Once more Roland found a way to attract an apparently heterogeneous and odd combination of participants: I found myself in a panel with John Lilly, an American expert on dolphins, and Jacques Mayol, the famous diver who was not yet the hero of the film *The Big Blue*.

The tragic death of Roland in 1988 passed unnoticed. Walk in the streets of a French city, and ask anyone at random who Roland Chevriot was. I doubt you'll be given the right answer ...

PREFACE

The following pages represent an attempt to study the history of the twentieth century. My point of view is practical. The focus is on two phenomena – industrialised farming and industrialised childbirth – which need to be analysed and understood in order to prepare for the forthcoming centuries. At a time when we begin to realise that there are limits to the domination of nature, even the role of history has to change. Historical books are still dominated by the relations, particularly conflicts, between human groups. Today the sudden evolution of the dichotomy between humanity and the laws of nature should become central to contemporary history.

After a century of industrialised farming and industrialised childbirth, reaching a new scientific truth implies the development of the capacity to think long term and see far into the future. Max Planck's comment in his *Scientific Autobiography* is more pertinent than ever: 'a new scientific truth does not triumph by convincing its opponents and making them see the light, but rather because its opponents eventually die, and a new generation grows up that is familiar with it'.

For this reason I hope this book will reach a new generation of young people.

ACKNOWLEDGEMENTS

I must acknowledge my indebtedness to all those whose advice and comments have influenced the content of this book, particularly: Sylvie Donna, Daniele Lamarche, Diane Speier, Jeannine Parvati Baker, Sara Wickham, the 'Queen of evidence-based midwifery' and Liliana, the 'Goddess of doulas'.

1 THE LAST STRAW

Epidemics used to be deemed natural disasters. Today that is not necessarily true. The turning point came when the twin scourges of foot-and-mouth and mad cow disease ravaged Britain and other European countries. Such epidemics have suddenly rallied public opinion against industrialised farming. They have opened the way to a new phase in the history of agriculture, animal breeding and husbandry. Foot-and-mouth disease was simply the last straw.

A EUREKA MOMENT

These events have given us the opportunity to realise just how deep the gap is between scientific knowledge and an awareness that leads to action. We have learnt that humanity as a whole can experience a kind of 'eureka' moment. Events inducing such a sudden awareness are unpredictable. Timing is the crucial factor.

The process of industrialisation tends to overpower and even to ignore the laws of nature, that is until the fateful day when spectacular disasters occur. We are now in a position to observe farming at its turning point. At the same time we are urged to look at other aspects of industrialisation, such as industrialised childbirth, which has direct consequences for human beings.

FOOT-AND-MOUTH

The case of foot-and-mouth disease is significant. According to Abigail Woods, a veterinarian and an expert in the history of the disease, 'Foot and mouth is as serious to animals as a bad flu is to human beings'. Until the end of the nineteenth century, it was common across the UK, and it is still endemic in many countries, including much of Africa, Asia and South America. Experts across the board furthermore claim that the disease has no implications for the human food chain and does not present a threat to public health. It affects cloven-hoofed animals. Human infection has been described but only a few cases have ever been confirmed by the isolation of the virus. No human-to-human transmission has ever been recorded. Nor has there been a survey to investigate the scale of silent human infection.

One must, of course, always be cautious with this family of viral diseases as the virus concerned has, in theory, potential to mutate rapidly. The rarity of the disease, despite long and close contact between human beings and infected animals, suggests however that the risk is remote. So, why such a fuss about one particular epidemic?

In the age of coloured pictures widely transmitted by the media, images of mass slaughter of healthy cows, pigs and sheep shocked a great part of the population, including farmers. This shock in itself was propitious to the advent of a new awareness. As Peter Melchett, a farmer in Norfolk, wrote in a British national newspaper: 'Many of us hope that foot-and-mouth will prove to be the last terrible gasp of intensive farming.' Timing, in fact, is the main reason why this epidemic will remain a landmark in the history of farming. Because it appeared at the time of mad cow disease, it became obvious that an instant-destruction policy is favoured by the entire system of industrialised farming. A disease that often reduces appetite and causes a drop in milk yield is incompatible with the concept of high productivity. Mad cow disease laid the foundation for a new awareness. Foot-and-mouth was the triggering factor.

MAD COW

Mad cow disease appears as the more worrisome of these two issues. The scientific term 'bovine spongiform encephalopathy' (BSE) clearly indicates that this infectious disease shoots the brain full of holes. It is one in a category of progressive neurological disorders called transmissible spongiform encephalopathies (TSE).

The fatal human nervous-system disorder, Creutzfeldt-Jakob disease (CJD), also belongs to this category. It is understandable that the panic began when a new variant of this terrible affliction was described and tentatively linked to mad cow disease. Today there is still a widespread belief that people can get the variant Creutzfeldt-Jacob disease (vCJD) from eating mad cows. While the real CJD mainly afflicts the elderly, vCJD appears among younger people. The quick and gruesome death starts with mood swings, numbness and uncontrolled body movements. Victims of the variant CJD usually die four months after symptoms appear. There is no treatment.

There are many reasons why the epidemics of mad cow disease created such widespread dread and fear. One reason is that the relationship between the cow disease and CJD remains unclear. The greater the mystery, the greater the fear. The mystery surrounding this issue is enhanced by the fact that the infectious agents causing these degenerative diseases of the nervous system are still surrounded by mystery. They are neither viruses nor microbes. They consist of protein and nothing else. They are 'prions'. The current dominant theory is that prions convert normal protein molecules into dangerous ones simply by inducing the benign molecules to change their shape. They are the toughest of all infectious agents, able to survive in conditions that would easily eliminate bacteria and viruses.

Another reason for anxiety is that in human beings TSE, with a 100 per cent fatality rate, is supposed to take many years to manifest symptoms. It will not be easy for some anxious beef-eaters to get rid of their sword of Damocles.

The anxiety induced by TSE is not simply localised to Europe. It is spreading to Japan, where at least one case has been observed. It is spreading to countries such as the USA and Canada, where neither mad cow nor vCJD have yet been seen. However most of the conditions thought to have led to the epidemics in Britain also exist on the North American continent. A similar epidemic in the USA would be even more catastrophic. Britain before the outbreak had roughly 10 million cows; in the USA there are more than 100 million. Avoiding an American outbreak of this disease may be down to only a matter of chance.

This latent state of anxiety was enhanced when it was revealed that certain drugs and vaccines, particularly polio, diphtheria and tetanus vaccines, have been made with products that could carry mad cow disease. What are, in reality, the risks of developing an illness such as vCJD from the use of bovine-derived material? A number of factors must be taken into account. These include the nature and the amount of bovine tissue used in the manufacturing process, as well as the date and country of origin of cows from which such materials are derived. The risk has been evaluated according to whether foetal calf serum from British origin, continental European beef broth, or bovine-derived materials from the USA have been used. According to the most pessimistic evaluations, when vaccinating 4 million children, or the entire birth cohort of the United States, risk levels would correspond to one case of vCJD arising every 5,000 years.

To date, nobody can seriously evaluate the risk of getting vCJD from eating mad cow. Manifestation of the condition possibly involves a genetic susceptibility that is not yet well understood. It has been claimed that patients seen so far with vCJD are probably genetically disposed to a short incubation period. We should therefore expect many more cases. In fact, it appears today that the number of cases is very much less than would be expected from a foodborne source.

At the same time we know that, according to tests on sheep and goats naturally infected with clinical scrapie, the most effective routes for the transmission of TSEs are those that require direct injection. The most effective one is intra cerebral. The least effective route is considered to be that of oral ingestion. We also know that, on a gram for gram basis, sheep brain is 100 million times more infectious than sheep muscle. We can therefore assume that it is the same with bovines. Thus, if the bovine brain contains 10 million infective doses per gram, then we can estimate that a beefsteak contains 0.1 infective dose per gram. It is difficult, yet necessary, to think in terms of order of magnitude.

A TIDAL WAVE

The space occupied in the media for more than a decade by mad cow disease is probably disproportionate to the real threat it represents for the health of humanity, particularly the unborn generations. But a widespread state of anxiety has been instrumental in creating conditions for a new awareness.

This sudden awareness was universally directed towards industrialised farming. Industrialised farming was the un-ambiguous target of even the most official reports. A British report to concerned British ministers could conclude that 'BSE developed into an epidemic as a consequence of an intensive farming practice – the recycling of animal protein in ruminant feed. This practice, unchallenged over decades, proved a recipe for disaster.' Soon after, the foot and mouth epidemic suddenly rose. The time was ripe for turning another page in the history of farming.

Such an explosive awareness had the effect of a tidal wave. While butchers remained sullen, supermarkets witnessed an overnight rise in pasta, rice and fish sales. In the media and in private conversations a multitude of issues surrounding industrialised farming were reconsidered in one go. New items about all breeds of farming livestock, from cattle and pigs to sheep, and horses, suddenly appeared in the press. The spectacular experiments by Albert Howard at the beginning of the twentieth century, in India, were suddenly saved from oblivion. Howard introduced a regime aimed at promoting the general health and well being of several different groups of oxen. He exposed these healthy animals to a range of diseases, all of which were endemic in India at that time. The oxen were brought into direct, nose-to-nose contact with animals that had foot-and-mouth disease. Since no infection took place, Howard could conclude that 'foot and mouth outbreaks are a sure sign of bad farming'.

In the context of the year 2001, when it was revealed that a mystery illness had killed hundreds of foals throughout Kentucky's Bluegrass country, causing huge financial losses in the world capital of racehorse breeding, an expert in equine epidemiology was quick to stress that the disease had nothing to do with BSE or foot-and-mouth. Articles about poultry farming and poultry diseases became prolific. Aspects of crop farming, vegetable farming and fruit farming became topical. Tongues were loosened about threats associated with pesticides, herbicides, fungicides and fertilisers. The watchword was: 'Eat organic.'

2 MAJOR PREOCCUPATIONS AND LATEST SPECTACULAR EVENTS

After referring to the epidemics and the spectacular animal slaughters that are at the root of a new phase in the history of farming, we must keep in mind what our main preoccupations should be. In an age of powerful means of communicating information it is commonplace to focus on the latest spectacular events. This explains our incapacity to establish a valuable classification of our main preoccupations.

In terms of public health, one of our main preoccupations should be intra-uterine pollution by a great variety of fat-soluble synthetic chemicals. Many of these chemicals are closely related to industrialised farming, particularly to the control of insects and to weed control. Whatever the chemical family they belong to, they are fat-soluble. Many of them have a long life. They accumulate over the years in our adipose tissues. We all have in our body hundreds of man-made synthetic substances, mostly polychlorinated chemicals, which would not have been there 50 years ago, because they did not exist at that time.

In order to anticipate the importance of the issue of intra-uterine pollution, we must first recall the main advances of the past 15 years regarding our understanding of health and disease. An overview of the Primal Health Research data bank can convince anyone that our health is to a great extent shaped in the womb (visit www.birthworks.org/primalhealth). There are in our data bank hundreds of studies detecting links between a state of health in adulthood, adolescence or childhood and what happened when the baby was still in the womb. Such studies are found in all fields of medicine and health sciences. They are unrelated according to the current classifications and, for that reason, difficult to detect in the medical and scientific literature. Some of them indicate that intra-uterine pollution has multiple long-term consequences. They even suggest that intra-uterine pollution with chemicals related to industrialised farming is a major threat to the health of the yet unconceived generations. The warnings must be taken seriously because they originate from a great diversity of medical disciplines.

NEUROLOGICAL AND INTELLECTUAL DEVELOPMENT

It was indirectly that we originally heard about the effects of intra-uterine pollution on neurological and intellectual development. It was through studies whose primary objectives were in fact to evaluate the effects of human milk pollution. This sort of pollution can be easily evaluated and is therefore well documented. Because all the widespread synthetic chemicals are fat-soluble, their presence in milk is predictable. On the other hand formula milk lipids are replaced by lipids of vegetable origin with a negligible content in PCBs and other polychlorinated chemicals. For that reason the inescapable first question is: do the well-known benefits of breastfeeding outweigh the theoretical risks associated with exposure to PCBs and other chemicals?

Dutch researchers have been particularly instrumental in providing answers to this simple question. A first Dutch study, published in 1995, looked at the neurological development of 418 children at the age of 18 months. Half of them were breastfed (for at least 6 months) and half of them were formula-fed. PCBs concentrations in cord and maternal plasma were used as a measure of exposure before birth. To evaluate postnatal exposure, the chemicals were measured in human milk and in formula milk (in formula milk it was 'below detection limit'). After taking into account many associated factors, it appeared that it was exposure to PCBs before birth that had a negative influence on the neurological condition at 18 months. On the other hand no negative effects of exposure to chemicals through breast milk could be detected. On the contrary, breast milk had a significant positive effect on the fluency of movements. The results of such a study suggest that, where neurological development is concerned, intra-uterine pollution probably represents a more serious threat for the unborn generations than milk pollution.

Another Dutch study, published in 1996, had many points in common with the previous one and led to similar conclusions: the benefits of breastfeeding still outweigh the negative effects of milk pollution; the focus should be on exposure to chemicals at an early phase in the development of human beings, that is before birth. Since 1996, year-by-year, the study has been updated with longer follow-up at 42 months and 6 years. The main conclusions remain the same.

The Dutch data are confirmed by the results of an authoritative American study of the intellectual functions of 11-year-old children. The authors originally recruited 212 babies born to mothers who had eaten Lake Michigan fish contaminated with PCBs. The presence of these PCBs is undoubtedly the consequence of human activities, particularly intensive farming. Concentrations of PCBs in maternal blood and milk at delivery were slightly higher than in the general population. Exposure in the womb was evaluated by measuring concentrations in umbilical-cord blood and by taking into account maternal blood and milk concentrations. When the children were 11 years of age, a battery of IQ and achievement tests was administered. Prenatal exposure to PCBs was associated with lower IQ scores after taking into account such factors as socio-economic status. The strongest effects related to memory and attention. The most highly exposed children were three times as likely to have low average scores and twice as likely to be at least two years behind in reading comprehension deficits in association with exposure before birth. Although larger quantities of PCBs are transferred by breastfeeding than through the placenta, there were only deficits in association with pollution before birth.

TOOTH DEVELOPMENT

Teeth are highly vulnerable at their stage of formation during foetal life.

Since the early 1980s, dentists from Finland have been studying how polychlorinated chemicals interfere with tooth development. They noticed that many children had poorly developed molars, discoloured and soft. The normal hard enamel coating was missing, making the teeth subject to decay. They took into account the effects of an accidental exposure to chemicals in Taiwan. A whole population had consumed cooking oils contaminated with polychlorinated chemicals. Children whose mothers were exposed while pregnant showed tooth problems similar to those of the Finnish children. Taking this as a clue, the Finnish dentists demonstrated that experimental exposure of rats to such chemicals leads to developmental defects of dental hard tissues.

To find out whether teeth could be used as a biomarker of exposure to polychlorinated chemicals, they examined dentitions of 102 children aged 6–7 for the presence of hypomineralised enamel defects. The permanent first molars were the target teeth. Severity varied from chalky lesions to localised loss of enamel associated with affected dentin. Mineralisation defects occurred more often and were more severe in children who had been exposed to higher amount of polychlorinated chemicals than in those exposed to lower amounts.

This study suggests that hypomineralised dental defects may be the best available indicator of prenatal exposure to chemicals, because defects are seen after exposure to very low concentrations and because such defects can be diagnosed even after many years.

THE MALE GENITAL TRACT IN DANGER

Scientists have recently understood that many man-made chemicals, particularly those transmitted via insecticides, pesticides, herbicides, fungicides and fertilisers, potentiate each other and mimic hormones. More precisely they mimic oestrogens, that is female hormones. That is why the male genital tract is in danger. That is why disorders of the male genital tract are increasing. No other plausible explanation has been offered so far.

Similar reports from various industrialised countries indicate that more and more boys have undescended testicles. A Spanish study compared the rate of undescended testicles in the different regions of the province of Grenada. Since this birth defect is typically corrected surgically, it was easy to calculate the frequency of the specific operation. Fruit and vegetable crops in the province of Grenada are treated with 51 per cent of the pesticides used in Spain. In much of the area along the Mediterranean coast, greenhouse crop farming under plastic-encased systems is widespread. In the enclosed greenhouses, workers (including pregnant women) are exposed to high levels of pesticides. It was found that the number of specific operations was significantly higher in districts where pesticide use was high.

Abnormalities of the penis, such as hypospadias – when the opening of the urethra is in the underside of the penis – are also more frequent. A recent analysis in the United States showed that the rate of hypospadias had nearly doubled in all four regions of the United States from 1970 to 1993. During the same period, testicular cancer rates have also increased. Today it is commonly accepted that most cancers of the testicles are the long-term effects of developmental defects before birth.

The spectacular fall of the average sperm count and the decline of semen quality since the middle of the twentieth century represents the most intriguing sign of the increased vulnerability of the male genital tract.

If all these chemicals are 'oestrogen-mimickers', it is likely that they can also, in a more subtle way, interfere with the development of the female reproductive system. This is suggested by the results of a Belgian study of sexual precocity after immigration from developing countries. The researchers

noticed that among 145 children seen for treatment of preco-
cious puberty, as many as 39 girls had immigrated 4 to 5 years
earlier from 22 developing countries, without any link to a
particular ethnic or country background. All these girls had
high blood levels of DDE, a chemical derived from the organochlorine DDT, which is still in use in the developing world.

MALE FOETUSES IN DANGER

Not only is the male genital tract in danger, but also the very
life of male foetuses is more in danger than ever. Of course
male embryos have always been more at risk of dying than
female embryos. According to an evaluation by T. Hassold,
in 1983, there were some decades ago 132 abortions of males
for 100 abortions of females. This implies that if the rate of
miscarriages is increasing, the ratio of male to female births
should be decreasing. This is exactly what is happening
today.

Recent reports from industrial countries indicate that the
proportion of males has declined significantly in the past
three decades. Among newborns in Denmark and the
Netherlands the proportion of males has declined in a
parallel manner between the 1950s and the 1990s. There
were similar trends in Canada and the USA for the period
1970 to 1990. For Canada, during this period, there was a
loss of 2.2 male births per 1000 live births. In the USA, there
was a decrease of 1.0 male birth per 1000 live births. It has
been observed that in some Latin American countries the
male proportion has also declined since the 1970s. Similar
trends have been reported in Finland and in Italy. It is highly
probable that prenatal pollution is the main cause. This

interpretation is supported by the reports following the 1976 industrial accident at Seveso, Italy, which produced the highest documented community exposures to polychlorinated chemicals. Between 1977 and 1984, 48 girls but only 26 boys were born to parents involved in the accident.

Until recently the increasing number of male foetal losses by miscarriage was the basis of a hypothesis explaining changes in the proportions observed at birth. A report of statistics from Japan has brought indisputable proof. The number of miscarriages in Japan was counted after 12 weeks of gestation, when it is possible to identify the sex of most foetuses. The proportion of male to female losses between 12 and 15 weeks gestation increased from 2.52 in 1966, to 3.10 in 1976, to 6.19 in 1986, to 10.01 in 1996! In other words, in 1996 in Japan, in this particular age group, ten boys were eliminated for one girl! Intra-uterine pollution is the only possible explanation for such differences.

LEARNING FROM FARMERS

Some farmers are highly exposed to insecticides, pesticides, herbicides, fungicides and fertilisers, and therefore to a great variety of fat-soluble chemicals. In fact, whatever our occupation, we all have in our body hundreds of such contaminants. The effects are just easier to detect when the levels of exposures are higher than average. That is why studies of farmers are particularly useful and must be taken seriously.

The results of a recently published study from Montreal about childhood leukaemia is a reminder that whatever the importance of the intra-uterine environment, we must realise that men also influence the health of the unconceived generations. According to this report, when a man is more exposed than most men to fat-soluble chemicals, his children are at increased risk of acute lymphoblastic leukaemia, which is the most common form of cancer in childhood. The results are statistically very significant regarding pesticides in general, fungicides and fertilisers. We must give credence to the Montreal study, which offers a spectacular example of what is called 'male-mediated developmental toxicity'. Today the usually prenatal origin of this type of childhood cancer is confirmed from a great variety of perspectives.

Not only can paternal pesticide exposure increase the risk of developmental defect, but it also can reduce the fertilising ability of couples. A questionnaire on occupational and life-style factors was completed by 652 Dutch couples that sought in-vitro fertilisation treatment between 1991 and 1998: 16 men were classified as occupationally exposed to pesticides. The association between pesticide exposure and fertilisation rate remained statistically very significant after taking into account many factors such as smoking habits, caffeine use, alcohol consumption and other occupational exposures.

This review of the direct health effects of synthetic molecules indicates that when we raise the issue of pollution and health, the focus should be on the health of the unconceived generations. From a practical point of view it also indicates that today our programmes of preconceptional preparation should be based on radically new preoccupations specific to our time. In traditional societies the main preoccupation before conceiving a baby was to create the best possible conditions to welcome a soul. In the 1980s the focus was on mineral imbalances and pollution by heavy metals. In the 1990s we learnt that folic acid supplementation can reduce the risk of abnormalities such as spina bifida. In the 2000s, we are prompted to start from novel questions such as: 'Is it possible to renew the adipose tissues of the body before conceiving a baby?' Our 'accordion method' is an attempt to adapt to these new preoccupations.

After mixing the events which are the root of a recent global Aha! and analysing the main threats for human health, we are in a position to exercise our human capacity for analogical reasoning. We feel ready to look at another aspect of the process of industrialisation – that is industrialised childbirth – with new eyes. We have the necessary basis for analogical reasoning. I find it relevant to take advantage of a jargon fashionable among experts in computational models of human intelligence. Now we have got 'the source' (industrialised farming); let us look at the 'target' (industrialised childbirth).

3 THE SOURCE AND THE TARGET

THE MOST FAMOUS EXAMPLE

The most famous eureka moment in the history of mankind was the effect of analogical reasoning. King Heiron had asked Archimedes to solve a problem. The king had given to a craftsman the required amount of gold to make a new crown. Although the finished crown was the same weight as the gold he had given originally, the king suspected that the craftsman substituted silver inside the crown and kept some of the more valuable gold for himself. Of course the king did not want to cut into such a beautiful crown. He wanted to find an indirect way to solve the problem. Archimedes promised to think about it.

Some days later Archimedes was relaxing at the public bath-house. As he sat in the tub he observed the amount of water displaced by his body. He noticed that the water level rose and fell as he lowered and lifted his weight ... An idea struck like lightning. He was so excited about his discovery that he jumped up and ran down the streets shouting 'Eureka, eureka!'. He had understood that if the crown displaced more water than expected, that meant that the crown was filled with something less dense than gold.

All the conditions for an Aha! had been ideally associated. The timing of the second event was perfect: having a bath after meeting the king. Archimedes was probably in an emotional state dominated by the fear of failing to solve the problem. Furthermore, analogical reasoning was involved.

A HUMAN TRAIT

It is easy to analyse and to explain the process of analogy. There is first a 'source': it is a piece of knowledge one is familiar with. Today industrialised farming can be used as a source. Everywhere in the world we all hear about the different aspects of modern farming. The limitations of industrialised farming have reached the phase of collective awareness.

In the process of analogy there is also a 'target'. This is usually a less familiar and more complex piece of knowledge, which has many similarities with the source. By observing similarity between the source and the target we can improve our knowledge and our understanding of the latter. We can also facilitate the emergence of a new awareness. Our target is industrialised childbirth. Industrialised farming and industrialised childbirth have many points in common. One can even claim that they are two aspects of the same phenomenon. In one case it is about non-human living creatures. In the second case it is about human beings. Both are typical ways to deviate from the laws of nature.

Analogy is a basic human reasoning process used in science, education, politics, literature, art ... It is widely used by orators and preachers, who need to convince their public. Analogical perception is a necessary ingredient for creativity. It is a cross-cultural human trait. Recent research has shown that even three-year-olds can solve analogies. A new awareness always implies analogical reasoning.

REFUELLING THE CASCADE

The impressive eureka moment humanity has recently experienced was the effect of a cascade of analogies. It started with a very small number of new facts. We learnt that the triggering factor for the emerging BSE epidemics in 1986 was traced to a food supplement that included meat and bone meal from dead sheep. From that time the effect of all news regarding animal breeding or agriculture was to initiate another stage in the cascade of analogies. This is how, for example, we can explain the media storm over genetically modified foods. This is how we suddenly all discovered new aspects of the issue of fish farming. This is how we realised that chemicals can be put into the hives by beekeepers, either to temporarily drive the bees from their shelter during harvesting operations, or for the prevention of bacterial diseases, or to treat the mites ...

Today we must not slow down the rhythm of this cascade. We must make sure that it keeps going. Our duty is even to refuel it and to extend it. The future of humanity is at stake.

4   SIMILARITIES

A quick allusion to history is the easiest way to highlight the similarities between industrialised farming and industrialised childbirth. Both phenomena developed side by side during the twentieth century. It is as if the domination of nature, which has been the basis of our civilisations for many millennia, had suddenly reached another order of magnitude. A threshold has been crossed.

ANCIENT GOALS AND RUDIMENTARY TOOLS

The ancestors of modern farmers were already sharing the goals of intensification when they were simply clearing away competing vegetation so that the sun could reach the plants they wanted to cultivate, or when they were protecting their plants and animals by guarding and fencing, or when they were learning to use such tools as axes, pointed sticks, ropes and knives. There have been determinant steps in the process of intensification, such as the use of oxen for ploughing. But what we call industrialised farming is a twentieth-century phenomenon.

In the same way all known human societies had reasons to interfere with birth physiology since time immemorial. The most rudimentary tools have been beliefs and rituals. These tools are particularly effective at disturbing the phase of labour between the birth of the baby and the delivery of the placenta. For example the cross-cultural belief that colostrum is tainted or harmful implies that the newborn baby must be immediately taken away from the mother. The effect is to challenge the mammalian maternal protective and aggressive instinct. Such a belief cannot be dissociated from a ritual, which is to rush to cut the cord before the delivery of the placenta. This is usually the role of a birth attendant.

The concept of the birth attendant is probably more recent than one commonly believes. Films among the Eipos in New Guinea and written documents about pre-agricultural societies suggest that there was a phase in the history of humanity when women used to isolate themselves when giving birth. Even more than for the other apes, privacy is a basic need for women giving birth. One must keep in mind the primary handicap of human beings in such circumstances. It is the enormous development of a part of the brain (the neocortex) that tends to inhibit the activity of more primitive brain structures. When one feels observed the neocortex (the brain of the intellect) cannot take a back seat. It probably occurred occasionally that a young woman in the bush was calling her mummy for help at the last minute. This is the root of midwifery. A midwife is originally a mother-figure. The advent of the birth attendant induced a vicious circle. The birth attendant interferes with the need for privacy and therefore tends to make the birth more difficult, so that there is a need for more help. In many societies birth attendants have used traditional methods to actively influence the progress of labour: manipulating, kneading, even bouncing on the abdomen or dilating the cervix manually. The widespread tendency to deny the mammalian need for privacy led to the socialisation of childbirth. The invention of tools such as forceps and the introduction of the medical man in the birthing place have been determinant steps in the history of childbirth. But what we call industrialised child-birth is also a twentieth-century phenomenon.

EXPLOSIVE DEVELOPMENT OF INDUSTRIALISED FARMING

By the early 1900s, the beginning of the explosive develop-ment of industrialised farming became visible.

What will become the symbol of industrialised farming lasted in fact a whole century. Swift and Company, the Chicago meatpacker, was the first to mass-manufacture rendered protein and fat as animal feed as early as a century ago. The use of such feed really took off in countries such as Britain during the Second World War, when it became difficult to import vegetable feeds from abroad. In Britain cows were fed as much as four pounds of rendered protein a day. By the mid-1970s the price of vegetable protein in the United States had risen to the point that rendered animal protein became an increasingly popular cattle feed – though never as popular as it was in Europe.

The progress of industrialised farming has been associated with a series of spectacular technological advances. Huge investments became necessary in order to take advantage of labour-saving machinery. This explains the need to concen-trate such facilities as tractors, combine harvesters and crop processing machines in larger and larger farming units. Intensive farming was instrumental in the development of the chemical and pharmaceutical industries, with the use of synthetic fertilisers, weed killers, insecticides, and also the treatment of animals with hormones, antibiotics and chemi-cals. One of the most typical aspects of industrialised farming is to 'scientifically' and electronically feed the cows. Each cow's ration is determined using a formula that takes into account the animal's weight, age, and milk output. This information is coded and integrated into a computer chip embedded in a tag that the cow wears on a necklace. A computer can read the tag, decipher its message, and dump precisely the amount and kind of feed the cow must receive.

EXPLOSIVE DEVELOPMENT OF INDUSTRIALISED CHILDBIRTH

It was also by the early 1900s that the fast development of industrialised childbirth became visible. It was originally more visible in the USA than in Europe. On both sides of the Atlantic the primary phenomenon was the increased control of the birth process by doctors. In a country such as Britain, the Midwives Act, which received in 1902 the Royal Assent after historical controversies in medical journals and endless negotiations with the General Medical Council, established official links between midwifery and the medical profession. It institutionalised a subservient role of the midwife to the physician. At that time, in America, the doctors had already gained a grip on the birth process and the status and role of the midwife were dwindling with equal rapidity. Midwifery was already associated with the so-called ignorant, illiterate immigrant women. There was no question of providing them with adequate training. While the major impetus for the elimination of midwives was disguised in terms of better care, there were also economic reasons. Not only was the volume of business for physicians limited by midwives, but since midwives' clients were mostly poor, the 'material' with which to train new generations of obstetricians was diminished as well. In such a context hospitalisation became widespread earlier than in Europe.

An American professor of obstetrics, Joseph DeLee, played a prominent role in the advent of industrialised childbirth. He was the author of several obstetrical textbooks, a sought-after speaker, and the inventor or modifier of many obstetrical tools. In his famous 1920 article and speech to fellow obstetricians entitled 'The prophylactic use of forceps', he noted that 'labor is a pathological process'. He recommended the routine use of forceps and episiotomy at every birth. He suggested that the 'patient' should be sedated and that ether should be given when the foetus entered the birth canal; ergot or a similar agent would be used to hasten the delivery of the placenta, which would then be extracted with a 'shoehorn maneuver'. DeLee's treatise was so influential in America that by the 1930s 'prophylactic obstetrics' had become the norm.

It was also around the turn of the century that work was begun in Germany on the effects of a mixture of drugs, morphine and scopolamine. Several American women were so excited by the prospect of a completely painless delivery that they went to Freiburg, Germany, during the opening months of the First World War. They returned to promote 'twilight sleep'. The technique involved first injecting the woman with morphine at the beginning of labour and then giving her a dose of the amnesiac drug scopolamine, which caused her to forget what was happening. During the second stage the doctor gave ether or chloroform. The campaign for twilight sleep was so successful that it attracted women to the hospital and at the same time made them more manageable during labour and delivery and allowed use of other techniques. With the advent of twilight sleep, hospital births became more impersonal. Staff, believing that women under scopolamine would remember nothing, had a tendency to ignore their 'patients'. On the other hand the making of all births as predictably similar as possible may have appealed to women who wished a speedy and painless delivery. Birth became an assembly line. The concept of prophylactic obstetrics promoted by Joseph DeLee, associated with the popularity of twilight sleep, explain why industrialised childbirth was well established long before the Second World War, at least in America.

The industrialisation of childbirth entered a new phase in the middle of the twentieth century. This was the effect of a series of technical and technological advances. In the 1950s the modern technique of 'low segmental' caesarean section gradually replaced the classical one. The principle of the new technique was to make the uterine incision horizontal, just above the cervix, at the level of the so-called 'low segment', which appears and develops at the end of pregnancy: the uterine incision was previously vertical in the main body of the uterus. This new surgical technique, associated with the advent of post-war methods of anaesthesiology, the organisation of blood transfusion, the availability of plastic material making drips safe and the possible use of antibiotics, suddenly transformed the C-section into a reliable operation. We must keep in mind that a C-section was such a dangerous operation at the beginning of the century that in 1910 its rate in America was around 0.2 per cent.

Such spectacular advances occurred at a time when most doctors involved in childbirth had no surgical background. Their main tools were forceps to extract the baby and scissors to cut the perineum. When a doctor thought that a caesarean was necessary as a last resort, it was usual to call a surgeon for the operation. This is a reason why, in spite of the advent of safe new techniques, the rate of caesareans did not increase dramatically until the 1960s when a new generation of surgically trained obstetricians appeared. The rate was still around 5 per cent in the USA in 1968, and still lower in Europe. During this phase of transition births were more and more concentrated in hospitals. It was easy to convince anyone that the best way to take advantage of the recent medical advances was to give birth as close as possible to an operating room, that is in a hospital with surgical facilities.

In the 1970s, at the very time when hospital births had become the norm and when a great number of autonomous and surgically trained obstetricians were practising, electronic foetal monitors suddenly appeared in the labour rooms. Within a few years the electronic age of childbirth was established. Instead of listening to the baby's heartbeats now and then it became routine to record continuously the rhythm of the heartbeats on a graph, thanks to an electronic machine. There was an increasingly complex network of tubes and wires around the woman in labour, including the tube of a drip connecting her arm to a bottle providing a calculated amount of synthetic oxytocin, the hormone necessary for uterine contractions. The atmosphere of the labour rooms was radically transformed. Women were giving birth in an electronic environment.

At the end of the twentieth century, at the height of the electronic age, the history of industrialised childbirth took another step. Epidural anaesthesia developed. It became clear that an epidural is the most effective form of obstetric analgesia. It is administered by injecting a numbing drug between two segments of the backbone through a fine plastic tube that is left in place in case more anaesthesia is needed. The technique was well codified before 1980, but at that time there were practical obstacles to its widespread use. Anaesthesiologists had been originally trained to make surgical operations possible, but not to intervene during a physiological process. Furthermore the practice of epidural anaesthesia is so time-consuming at unpredictable hours of the day or night that very few hospitals could offer a round-the-clock service. That is why the fast development of epidural anaesthesia had to wait until the 1990s, when a sufficient number of specialised practitioners were available, and also when more sophisticated techniques of 'walking epidural' had been evaluated.

At the dawn of the twenty-first century only gigantic maternity units can offer a service based on the presence of obstetricians, anaesthesiologists and paediatricians 24 hours a day. That is why, in Western Europe, the average number of births per day in many hospitals is 10 or more, while on the American continent it can be above 20. Huge investments have often been necessary to design, build and equip large enough maternity units. Although our focus is on the birth itself, it is clear that these huge investments must also cover the costs of the sophisticated equipment necessary for modern prenatal care, particularly ultra-sound machines.

Concentration in large hospitals is not the only characteristic of industrialised childbirth. There is also a striking tendency towards standardisation. 'Routine' and 'protocols' are key words in modern obstetrics. In the minds of many people, apart from birth by caesarean, which can be planned or decided during labour, there is a 'normal', quasi standardised birth. In the case of a 'normal' birth the woman is given

an epidural and a drip of oxytocin, while the baby is
electronically monitored. It is usual that a tube is passed up
from the urethra to drain the bladder. During the last
contractions the use of a ventouse (or forceps) is associated
with an episiotomy. At the very time when the baby is born
a drug is routinely given in order to contract the uterus for a
safe delivery of the placenta. In the age of industrialised
childbirth the mother has nothing to do. She is a 'patient'.

5   ENTHUSIASM

There are other similarities between industrialised farming and industrialised childbirth. One of them is that every new episode of their short history has been enthusiastically welcomed.

ENTHUSIASTIC FARMERS AND ECONOMISTS

The collective enthusiasm for industrialised farming was understandable. We must keep in mind that today the planetary resources must feed six billion human beings. This is deemed impossible with old-fashioned farming methods.

Farmers have expressed their enthusiasm at different stages of the process of industrialisation. In the past farmers regarded workers' wages as one of their heaviest financial burdens, so it was natural that they would enthusiastically look to machinery. Let us think of farmers raising cotton, for example. In the recent past their main expense was for weed control by hand labour. Herbicides have overnight dramatically reduced the costs. We can make similar comments regarding other aspects of agriculture in relation to the introduction of cheap and effective insecticides, fungicides and fertilisers. The benefits of industrialisation appear still more spectacular when considering particular aspects of animal breeding. Listen to a not-so-old American farmer who remembers that when he became a dairyman a good milking cow gave about 35 pounds of milk a day. Today his top 'milkers' give up to 130 pounds of milk a day thanks to protein supplements that can be made from almost any organic material, including corn, soybeans and rendered animal by-products.

The advantages of selective breeding are also obvious. This means, for example, that while some cows are bred to produce as much milk as possible, other animals are bred to produce as much meat as possible. Farmers learnt that by adding to the diets of growing animals antibiotics such as tetracycline or penicillin they could produce more meat more quickly and at the same time decrease death loss. The productivity is still better when the animals are treated with various combinations of hormones such as oestrogens, progesterone and testosterone. Many farmers remain convinced that factory farming, which includes housing in small compartments, allows them to tend their livestock carefully and that prices of farm produce would rise steeply if they were to give up the system.

Lay people also have serious reasons to be grateful for industrialised farming. We might multiply the examples. We must remember the time when the pelvis of many women was distorted by inadequate diets; for that reason childbirth presented many hazards to women. Not so long ago, most people could not have an ideal intake of vitamin C all the year round, a vitamin the human body cannot synthesise. Today, whatever the season, we can find lemons, oranges and other fruit rich in this essential vitamin on the shelves of our supermarket. Not so long ago smoked salmon was a luxury. Today it is within the reach of almost all family budgets. It is thanks to all the different aspects of intensive farming that we all have access to a great variety of food in our daily life. The health implications are enormous and easily ignored or underestimated. Food diversification is probably one of the main factors explaining the recent spectacular increase in life expectancy in wealthy countries and still more the increase in healthy life expectancy, that is the tendency to stay healthy to an older and older age. Diversification is also one of the main factors explaining the gradually increasing average height of men and women throughout the twentieth century.

During the twentieth century the public health benefits of industrialised farming have been enormous and probably underestimated. The cause and effect relationships are not always easy to demonstrate. However several phenomena, observed during the past two or three decades, are difficult to interpret without referring to the sudden access to a great variety of food.

A typical example is offered by a study about the fluctuations of the incidence of neural-tube defects, which are anomalies such as spina bifida. Everybody heard in the 1990s that the best way to reduce the risk of such defects is to take supplements of folic acid during the period surrounding conception. The most prestigious medical journals – the *Lancet* and the *New England Journal of Medicine* – published their convincing studies in 1991 and 1992, which were followed by intensive public health campaigns. However, according to data from the British Isles Network of Congenital Anomaly Registers, it was between 1980 and 1985 that there was a significant drop in the incidence of neural-tube defects. The public health campaigns had no effect on the graph, which remained perfectly flat after 1985. The question is: what happened between 1980 and 1985? This was the period when the European supermarkets developed. This means easy access to a great variety of food. There are strong links between industrialised food distribution and industrialised farming.

Twentieth-century human beings had multiple reasons to underline the advantages of industrialised farming.

ENTHUSIASTIC WOMEN AND ENTHUSIASTIC OBSTETRICIANS

There has always been a female determination to overcome the obstacles of childbirth, and a major obstacle to overcome at the beginning of the twentieth century was the fear of dying or being permanently injured. One must keep in mind that a century ago, in the USA, for example, the risk of dying from pregnancy or childbirth was above 400 per 100,000. That is why it was easy to convince women that the elimination of 'hopelessly dirty, ignorant and incompetent' midwives and the expansion of obstetrics constituted the solution and could provide relief and successful birth outcomes. That is also why the gradual switch of birth-place from home to hospital was welcomed by a great part of the population. For similar reasons, in Western Europe, new regulations increasing the control of midwifery by the medical profession were considered wise and relevant.

When Joseph DeLee recommended routine episiotomies and routine forceps at every birth, American doctors expressed their enthusiasm by making these emergency measures standard practice in maternity hospitals throughout the USA. These interventions were quickly accepted by women as well. Women who went to Germany in 1914 to have a twilight birth brought back their highly contagious enthusiasm. In the language of the 1930s twilight sleep 'streamlined maternity's miracle'.

As a surgeon educated in the 1950s I cannot forget the advent of the modern technique of low segmental caesarean section. It was undoubtedly the pivotal twentieth-century advance in the field of childbirth. What a wonderful rescue operation! I cannot forget the reactions of an old hardened surgical nurse who had participated in thousands of operations of all kinds. At the end of a caesarean she could not stop crying and repeating: 'this is the most beautiful operation I have ever seen!'

In the 1970s, the enthusiasm of many obstetricians for electronic foetal monitoring was in fact a real fascination. It started from a simple idea: let us record the baby's heartbeats continuously with an electronic machine, and we'll be in the best possible situation to immediately rescue a baby in danger. We'll bring more safety to the field of childbirth. This is how, within a few years, the electronic age of childbirth was established. This enthusiasm was reinforced by the publication of impressive statistics. In the 1980s the maternal mortality rates in Western Europe and America were in the region of 8 per 100,000, while, some decades before, they were calculated in terms of hundreds per 100,000. Another order of magnitude! It therefore became possible to change the focus and to look at the rates of babies' deaths. While the perinatal mortality rates were not long ago expressed in tens per thousand, in many wealthy countries they were below ten at the height of the electronic age. Obstetricians were quick to establish a cause and effect relationship between the use of electronic monitoring and improved statistics ... without waiting for the results of further studies.

We will not comment on the benefits of epidural anaesthesia. We only need to listen to the countless women who reported their experience with praise. A young woman proudly told me that she could watch the TV while giving birth. Another one could finish her crossword while waiting.

As for the enthusiasm inspired by the principle of planned caesarean section on demand, it is eloquently expressed by a Brazilian mother, whose comments were reported in the *Wall Street Journal*: 'I checked into the hospital ... like I was checking into an hotel ... Why go through all the anxiety, when you can arrange everything in advance?'

What can we add to so many reasons for enthusiasm!

6   REMEMBER THEM!

We underlined that the various steps of both industrialised farming and industrialised childbirth have been in general welcomed and considered highly beneficial. In spite of that we cannot ignore today the discordant viewpoints and warnings that have been expressed all through the twentieth century by a small number of outsiders. These outsiders were endowed with an exceptional capacity to realise beforehand the long-term consequences of human actions. They were able to think in terms of civilisation and not only in terms of individuals. They had enlarged visions. They can be presented as visionaries. They must be remembered by the future occupants of this planet.

VISIONARIES

Rudolf Steiner is historically one of the first of a series of visionaries. It is difficult to realise how the visions of one extraordinary human being, who died in 1925, are in tune with the problems of the twenty-first century. The influences of Steiner's insights are stronger than ever in many practical fields, encompassing arts, science, education, farming, medicine and social matters. I realised the particularities of the 'anthroposophical lifestyle' – that is a lifestyle influenced by the work of Steiner – when I was studying the possible links

between whooping cough vaccination and asthma in childhood. I found by chance some unexpected health effects. For
example, among the 210 pupils of a French Rudolf Steiner
school, aged 5 to 18, only 4 of them needed to wear glasses.
More recently a prestigious medical journal studied the low
rate of allergies among children who share an anthroposophical lifestyle.

Rudolf Steiner could not dissociate his interest in plant
development, animal development and human development.
This simple fact is a valuable lesson at a time when we are
the victims of the sort of blindness generated by narrow
specialisation. Some of the insights of Steiner may be
deemed incredible – even implausible – in retrospect, when
taking into account the scientific context in which they were
expressed. He was already concerned by the idea some
farmers had to feed cows with animal products. In a
conference in Dornach, on 13 January, 1923, he claimed
that if cows were given meat to eat, they would become mad!

The biodynamic movement sprang from eight lectures
that Rudolf Steiner gave in the early 1920s in response to a
request from a number of farmers. It was the first organised
alternative method of agriculture to be based on a comprehensive picture embracing both ecology and social life.
Biodynamic farming anticipated the destructive effects of
conventional farming: the soil will be eroded, the humus will
be lost, flowers and animals will disappear – damage which
will have to be suffered by future generations. The biodynamic movement was a powerful warning of the threat for
humanity of the cold-hearted exploitation of the resources of
the earth. It was a powerful warning because it was constructive, offering alternatives. Biodynamic farming involves
restoring to the soil a balanced living condition through the

application and use of the completely digested form of crude organic matter known as stabilised humus. Crop rotation, correct composting and proper intercropping can all contribute to a healthier biodynamic yield. The anthroposophical concept that pests and diseases are nature's way of getting rid of something that is basically unhealthy was in itself a warning. Today billions of dollars are spent on pesticides, fungicides and herbicides. However farmers still lose the same one-third of crops.

Robert McCarrison also had an 'intellectual passion for wholeness'. He was instrumental in developing and promoting the concept that the health of human beings, the health of animals, the health of crops and the health and fertility of the soil are inseparable. One of his revolutionary contributions to medicine was to shift attention from the prevention and the treatment of diseases to the development of good health. Another major contribution was to promote the concept that the health of the individual cannot be considered in isolation from the family, nor the health of the family in isolation from that of the wider community.

McCarrison, as a medical doctor, joined the Indian Medical Service in 1901. He stayed in India until 1935. His seven years of practice among the Hunzas were the critical years of his career. These tribesmen on the North West Frontier had none of the most common complaints in Europe. Among them McCarrison saw no heart disease, cancer, appendicitis, peptic ulcer, diabetes or multiple sclerosis. He tried to analyse the factors that could explain such good health. He took note of the food they were consuming in their daily life. He noticed that not only was human ill-health rare among them, but their farming was remarkably free of plant disease. This is how he became an observer of their farming methods. He found that the most important feature of Hunza cultivation was fidelity to the 'Rule of Return'. This implies a refusal to waste anything, which ensures that every possible material, including human waste, is composted and used to enrich the earth. He observed how they 'spread out the compost evenly like butter upon bread'.

The scientific mind of McCarrison led him to undertake animal experiments in order to complete what he was learning from clinical practice. He tested thousands of rats, feeding them on various faulty foods and comparing their health and longevity with a control group of well-fed stock rats. He even fed a group of rats on a Sikh diet, and another on a diet common to the poorer classes in England: white bread, sweet tea, boiled vegetables, tinned meat, jam and margarine. The well-fed rats flourished physically and co-existed harmoniously, compared with the others who suffered a greater incidence of disease, particularly pulmonary and gastro-intestinal.

By combining what he learnt from clinical practice, from the way the Hunzas obeyed natural law and from a great variety of animal experiments, McCarrison had understood the 'wheel of life', the cycle by which waste nourishes the soil, creating healthy plants, which creates healthy animals and humans, whose waste, properly treated, further nourishes the soil.

Wilheim Reich should also be remembered as one of the lucid giants of the twentieth century. His main field of interest was the nature of 'life energy'. This implies an interest in all aspects of life. Reich was one of those extraordinary men who are able to step outside their culture and examine it with innocent eyes. As early as the first half of the twentieth century he raised the most vital questions of our time. He wondered why humans fail to realise that they are parts of nature, and therefore must cooperate with it and obey its law. He studied the process of desertification. He came to the conclusion that it is the emotional desert in man which creates the desert in nature, referring to the huge capacity humans have to unhesitatingly destroy life. He understood that at the root of this widespread 'emotional desert' is the damage we do to newborn babies: 'Let's concentrate on the newborn ones and let's divert human attention away from evil politics and toward the child.' He was unambiguously hard on his contemporaries: the 'Children of the Future ... will have to clean up the mess of this twentieth century'. From his point of view civilisation will start when the well being of newborn babies will prevail over any other consideration.

The name of **Ina May Gaskin** is associated with 'The Farm' and with 'authentic midwifery'. Through her lifestyle, her acts and her teaching Ina May has transmitted essential and wide-ranging messages to her contemporaries. These messages can be easily condensed: humanity cannot survive without rediscovering the laws of nature; the first step, which should be to reconsider the way babies are born, implies the revival of authentic midwifery; another step should be, for the sake of the unborn generations, to stop destroying the soil through aggressive farming methods.

In 1971, 320 San Francisco hippies left the west coast. Their vision was to invent a new lifestyle. They crossed the country in a caravan of converted school buses. Their slogan was 'out to save the world'. They eventually created 'The Farm' community in the poorest county of Tennessee, near Summertown. During the period of transit 12 babies were born in the community. This is how Ina May and other mothers in the group became midwives.

When the community was settled, the members of the group had to gradually develop all of the usual implements of village life – grocery store, school, water systems, pharmacy, post office, cemetery, scores of businesses and residences – and birthing facilities. This is how Ina May and other mothers in the community learnt midwifery out of need and through experience, with the help of a sympathetic local doctor.

In the Farm heyday in the late 1970s two dozen babies a month were born in the community. Soon the midwives had acquired such experience that in 1977 Ina May was in a position to publish her historical book *Spiritual Midwifery*, a landmark in the history of childbirth. This book remains the symbol of the revival of midwifery in the USA, and also in Europe. I know several influential European midwives who visited the Farm after reading the book.

At the same time the Farm's founders, including Stephen, Ina May's husband, were experiencing organic farming. They gathered all organic waste from kitchens, stables, the canning and freezing plant, sawmill and grocery and transported it to a central composting facility where it was turned by horses and then spread on the fields. More than 300 acres were restored to productive organic agriculture. Large orchards and vineyards were planted, as well as fields of strawberries, raspberries and blueberries. A tree nursery was set up to propagate useful varieties of fruit trees and hard- and softwood. Terracing and engineered drainage halted soil erosion from the tillable lands. Apiaries were constructed to pollinate fields, nurseries and orchards. Use of polyculture and heritage seeds, cover-cropping, crop rotation, hand-picking, beneficial insects, snakes, lizards, toads and turtles provided organic pest control. Based on this experience, an organic pest control handbook has been published by the Farm Book Company.

The warnings by Ina May Gaskin and the other 'ecovillagers' of the Farm are powerful because they are supported by practical experience. They are not presented as purely negative critiques of industrialised farming, industrialised childbirth and other aspects of the dominant modern lifestyle.

Frederick Leboyer, the poet-obstetrician, has also been an influential visionary. In his famous book *Birth Without Violence* he suddenly invited us to share his vision of the experience of being born, while it was usual to write or to talk about giving birth. The timing was perfect. The book was originally published in French in 1974. This was already the height of the electronic age of childbirth. That is why 'birth without violence' was first and foremost perceived as a warning against industrialised childbirth. Not only the timing, but also the format of the book was ideal to induce a new awareness. There is a well-balanced association of words and significant, moving and beautiful pictures. The style itself also contributes to continuously stimulating the attention while maintaining an appropriate emotional state. It is the oracular style of a visionary, each paragraph containing only one, two or three sentences ... the style of holy scriptures.

Although Leboyer invites us to look at the newborn baby as

an individual, even as a 'person', it is clear that he is constantly thinking in terms of civilisation. The topic is enlarged by a small number of well-placed significant allusions. Leboyer does not miss the basic questions. While trying to interpret the ritual of rushing to cut the cord, he wonders: 'How is it that man, a rational animal of reputed intelligence, acts so irrationally in such an important moment?' One sentence, one question: the cultural dimension and the issue of aggressive rituals that welcome the newborn babies in all known societies are subtly introduced.

The way we are born explains how we behave at the time when a baby is born. According to Leboyer this process of 'transference' is endlessly repeated. 'And the sum of these repetitions is what we, in our ignorance, call education.' After being invited to enlarge our vision of how the cultural milieu interferes, and after being conditioned to think long term, we suddenly read:

> Such is birth. The torture of an innocent. One should have to be naïve indeed to believe that so great a cataclysm would not leave its mark. Its traces are everywhere; on the skin, in the bones, in the stomach, in the back, in all our human folly, in our madness, our tortures, our prisons, in legends, epics and myths. The scriptures themselves are surely none other than this abominable tale of woe.

This list of extraordinary human beings is not exhaustive. There have been other pioneers and visionaries who were able to clearly raise, during the twentieth century, the unavoidable questions of the third millennium. I have an understandable tendency to select those I am personally familiar with. This is the case with Ina May Gaskin and Frederick Leboyer, whom I often met. Because of my participation in meetings of the society named after him, this is indirectly so in the case of Robert McCarrison. Because of my friendship with his beloved daughter Eva, this is also indirectly true in the case of Wilheim Reich as well. As for Rudolf Steiner, he died before I was born. However I have a feeling of familiarity because of my strong links with people whose lifestyle and philosophy are inspired by the anthroposophical concepts.

7 NATURAL CHILDBIRTH AND ORGANIC FARMING MOVEMENTS

Bringing into focus a selection of influential gigantic visionaries should not overshadow the specific roles of countless active associations of campaigners in consciousness-raising. Such movements usually start when a small group of people are devoted to one particular cause. Their objectives are limited to one precise issue. Organic farming movements and natural childbirth movements have many similarities in this regard.

While visionaries anticipate questions and answers and tend to break down the partitions between conventional perspectives, movements tend to restrict their objectives in order to achieve specific goals as quickly as possible. Visionaries and movements are complementary. Directly or indirectly visionaries often inspire movements.

BIODYNAMIC AND ORGANIC FARMING MOVEMENTS

Throughout the twentieth century there have been movements promoting alternatives to industrialised farming. There have always been small regional grassroots associations. In the rural areas of a country like France, they play a major role in preserving a critical attitude among the local population regarding the different aspects of industrialised farming. The objectives of such associations are strictly limited to one particular geographical area. They are called, for example, Agro-bio Poitou-Charentes, or Allier-Bio, or Confederations des Groupes des Agrobiologistes de Bourgogne or Groupement des Agriculteurs biologiques de Touraine, etc. There are, on the other hand, a small number of large-scale organisations which work at national and international levels.

Historically, the first far-reaching organisation was undoubtedly the Biodynamics Farming and Gardening Association, which was formed in the US as early as 1938 in order to foster, guide and safeguard the biodynamic method of agriculture. There are several reasons why the biodynamic movement remained avant-garde for two-thirds of a century. The first reason is that it is primarily constructive, rather than being purely critical of the different aspects of industrialised farming, as they gradually developed during the twentieth century. We might say that the biodynamic movement has transcended the different phases of the history of industrialised farming, because its objectives are positive: learning to work with 'the health-giving forces of nature'.

The second reason for remaining avant-garde is that the Association has a great variety of activities. They include conferences, workshops, seminars and research. The association is the publisher of *Biodynamics*, America's oldest ecological farming and gardening magazine. It also plays an important role in advising farmers and gardeners. It supports regional grassroots associations and has its own database of CSA (Community Supported Agriculture) farms. It is associated with more formal research and training institutions, including the Josephine Porter Institute, which produces and distributes biodynamic preparations, and the Demeter Association, which certifies biodynamic farms. The biodynamic movement has branches in dozens of countries.

The organic farming movement began during the Second World War in the US. It was precisely during the war that agriculture in the US became highly industrialised using chemicals, mechanisation and monoculture to increase production rapidly and feed war-torn Europe. The organic movement appeared as an immediate and urgent reaction to the explosive development of industrialised farming. The new meaning of the word 'organic' was popularised as early as 1942, when Jerome Rodale launched the magazine *Organic Farming and Gardening* (which became *Organic Gardening*).

Immediately after the war, in 1946, in the UK, the Soil Association had its inaugural meeting. Eve Balfour's book *The Living Soil* was instrumental in drawing together those who founded the Association. Eve Balfour had met both Howard and McCarrison and was highly influenced by them. The association had a triple objective. The first one was to bring together all those working for a fuller understanding of the vital relationship between soil, plant, animal and man; the second one was to initiate, co-ordinate and assist research in this field. The third one was to collect and distribute the knowledge gained so as to create a body of informed public opinion. The third objective was the reason for the Association's journal *Mother Earth*. By 1953 membership had passed 3000 and was worldwide.

This post-war period was also the time when the US production of synthetic chemicals used by farmers increased exponentially. In 1960 US farmers sprayed approximately 300,000,000 pounds of synthetic pesticides on crops. They were gradually imitated by their European colleagues. The situation was ripe for the beginning of a new awareness among a small part of the population, including a limited number of farmers. Small grassroots associations multiplied. The rising reaction against industrialisation was suddenly reinforced in 1962, when *Silent Spring*, by Rachel Carson, appeared on bookstore shelves. It was an overnight sensation. This book offered the first shattering look at widespread ecological degradation. It focused on the poisons from insecticides, weed killers, and other common products as well as the use of sprays in agriculture, a practice that led to dangerous chemicals at the food source. Rachel Carson argued that those chemicals were more dangerous than radiation and that for the first time in history, humans were exposed to chemicals that stayed in their systems from birth to death. Presented with thorough documentation, the book opened more than a few eyes to the dangers of industrialised farming.

Around 1970, when there was a multitude of small associations all over the world, a group of clever and motivated pioneers felt the need to structure organic farming. The creation of the International Federation of Organic Agriculture Movements (IFOAM) was initiated by a small number of dedicated pragmatic friends such as Roland Chevriot, Mary Langman, Karin Mundt, Claude Aubert and Denis Bourgeois. One of their first objectives was to publish standards for organic agriculture. The urgent need for such an organisation was demonstrated by the spectacular development of IFOAM, which can be illustrated by eloquent figures. At its creation the federation had five member organisations. In 1984, at the general assembly in Witzen-hausen, Germany, there were about 100 member organisations from about 50 countries, representing a total of 100,000 individuals. At the beginning of the 1990s, one might have thought that the organic farming movement had reached its main objectives. In 1992 the European Union approved the first government-enforced standards for organic production. It was also the year when the world looked towards Brazil, where the United Nations Conference on Environment Development (UNCED) took place in Rio. There, IFOAM was very active in promoting the position of organic agriculture. IFOAM organised the first major international conference on environmental issues after the events in Rio with the Ninth International Scientific Conference taking place in Sao Paulo. It is at this conference that the UNO (United Nations of Organic) choir made its acclaimed public performance of the IFOAM anthem, using the melody of 'Auld Lang Syne':

> In all the world the need is felt
> To make a drastic change
> A choice for life, a choice for health
> Ever wider is the range
> So let us sing to living soil
> Organic farmers' pride
> IFOAM brings us all together
> To reach this goal worldwide.
> They herd the cows, they plant the seeds
> Not only humans do they feed
> Also water, soil and air
> So let us sing to living soil
> Organic farmers' pride

> IFOAM brings us together
> To reach this goal worldwide.
> May all our children and their children
> Live on a greener earth
> For their inherit all our deeds
> That is what makes it worth
> So let us sing to living soil
> Organic farmers' pride
> Join hands and may the work be blessed
> To reach this goal worldwide.

The words of the anthem clearly indicate that the main preoccupation of the representatives of IFOAM has always been the future of humanity in general. Their interest was not limited to their own health or the health of their family.

In spite of all these achievements, the mission of IFOAM was not over in the 1990s. Vigilance was imposed by new developments. These new developments explain the sudden emergence of the latest organic farming movement at the turn of the century.

The Keep Organic *Organic* Movement emerged abruptly in 1998, when the United States Department of Agriculture (USDA) revealed its plans to allow food that was genetically engineered, irradiated, or grown in sewage sludge to be called organic. The USDA was obliged to re-write in 1999 the standards after 275,000 citizens flooded them with comments concerning their initial attempt at regulating the organic label. The re-written rules introduced all the possible concessions to organic advocates and even kept antibiotics and growth hormones out of organic meat and dairy production. Accredited certifiers were allowed to uphold higher standards than the USDA.

The latest rules came at the very time when organic food sales were constantly growing. They are part and parcel of a new collective awareness. Today the priority is not to promote organic food any longer. It is to keep vigilant. The shift from industrialised to organic farming cannot be easy and cannot be realised overnight. For example it is worth noticing that according to the European Union regulations the label 'organic' is compatible with the use of copper sulphate to combat diseases such as potato blight, although it kills beneficial insects such as earthworms, and has caused liver damage in the vineyard workers. Other chemicals approved for use by European organic farmers include rotenone, the active constituent of the root of the Derris tree, which has been associated with Parkinson's disease. Today the priority is to keep 'organic' organic.

NATURAL CHILDBIRTH MOVEMENTS

Throughout the twentieth century there have also been movements reacting against the industrialisation of child-birth. The first reactions came from America in the 1920s; they were inspired by the practice of twilight sleep, which led to the concentration of births in hospitals and which made births more impersonal. *Twilight Sleep*, a satirical novel of the jazz age by Edith Wharton, published in 1927, was instrumental in helping a certain number of women to transmit their reluctance to share the dominant enthusiasm. Through one of her characters Edith Wharton clearly gave her point of view, with an unambiguous reference to industrialisation:

'Of course there ought to be no Pain ... nothing but Beauty. It ought to be one of the loveliest, most poetic things in the world to have a baby', Mrs Manford declared, in that bright efficient voice which made loveliness and poetry sound like the attributes of an advance industrialism, and babies something to be turned out in series like Fords.

It was mostly after the Second World War that countless associations appeared in industrialised countries. Some of them were small, with local or regional objectives; others had national or international dimensions ... another similarity with the reactions to industrialised farming. Each group has a history and a particular vocation. Even if the means they had proposed were not always appropriate and could even be counterproductive, all these groups shared the sane objective to seek alternatives to industrialised childbirth.

The prototype of a large well-organised association is undoubtedly the National Childbirth Trust (NCT) in the UK. It was launched in 1957 by mothers who had been highly influenced by the work of Grantly Dick-Read, the author of *Natural Childbirth*, published in 1933, and of *Childbirth without Fear*, published in 1944. The dominant idea at the root of NCT was that very little information about pregnancy and birth was available. The resulting ignorance bred fear, and fear led to pain. Over the years, NCT had to adapt to new situations. Today it presents itself as a centre for information about pregnancy, childbirth and lactation. NCT promotes informed choice, rather than one particular way to give birth. However the effect is to constantly recall that there are alternatives to industrialised childbirth. For example a recent survey on the net was in fact a subtle way of suggesting that in the context of the UK home birth is a safe option. Women were asked to answer yes or no to the following question: 'If your pregnancy is/was straightforward, was home birth presented to you as positively as hospital birth?' There is a link between the development of NCT and the degree of awareness which has been continuously nourished through the books by Sheila Kitzinger, an originator of NCT.

Still in the UK the Active Birth Movement has a different history. The reason for the phrase 'active birth', coined by Janet Balaskas, was to contrast with the phrase 'active management of labour', used originally in Dublin. Active management suggests that the labouring woman is not the active person. The Active Birth Movement is characterized by the existence of an Active Birth Centre in London. It has inspired the launching of groups with similar motivations in other countries, such as Nascita attiva in Italy and Naissance Active in Geneva. In Germany, the Gesellschaft fur Geburtsvorbereitung has similarities with both NCT and Active Birth. In France there have never been large national organisations comparable to the British ones, but a great number of small local groups.

The International Childbirth Education Association (ICEA), based in the US, has many similarities with NCT. It is an umbrella organisation of consumers and childbirth educators that promotes freedom of choice based on knowledge of alternatives in birth. It can really be considered international, since there are members in 42 countries. In the US, most groups look at childbirth from specific perspectives and have specific primary objectives. For example the International Cesarean Awareness Network (ICAN), which was preceded by the Cesarean Prevention Movement (CPM), Informed Homebirth, New Nativity, the American College of Home Obstetrics, the Maternity Center Association or the National Association of Childbearing Centers have transparent priorities. The National Association of Parents and Professionals for Safe Alternatives in Childbirth (NAPSAC), founded in 1975 by Lee and David Stewart, had a well-recognised vocation for encouraging home birth, although this is not its declared specificity. In the same way the American Foundation for Maternal and Child Health and its founder Doris Haire have a strong reputation for establishing and reinforcing links between groups and persons in different parts of the world who might ignore each other. Doris Haire might be presented as the Roland Chevriot of natural childbirth.

Certain American groups have a specificity that is expressed in terms of philosophy. This is the case of Birthworks, created by Cathy Daub. It embodies the philosophy of developing a woman's self-confidence, truth and faith in her ability to give birth through education, introspection and confident action. I intentionally do not mention groups whose objectives are to promote 'methods of childbirth'. Any 'method' is easily integrated into industrialised childbirth.

The natural childbirth movement also includes groups that have played such specific roles that they cannot be listed in our classification. For example the Australian Associates in Childbirth Education (ACE), created by Andrea Robertson, have acquired a unique reputation in organising workshops, seminars and conferences for health professionals. I know from experience that through a series of study days in the main Australian cities organised by ACE it is possible to reach nearly one thousand midwives ... an impressive number when considering the population of Australia. The Association for Improvement of Maternity Services (AIMS) and its founder Beverley Lawrence Beech, in the UK, plays a unique role in tracing and making public the results of scientific studies that challenge the very foundations of industrialised childbirth.

The natural childbirth movement would be hopeless without the activity of groups whose mission is to prepare the revival of midwifery. The revival of midwifery is the prerequisite for entering the post-industrialised era of childbirth. That is why we must acknowledge the paramount importance of such groups as Il Marsupio in Italy, which created an alternative midwifery school; Spiritual Midwifery in Russia; The Associations of Radical Midwives in the UK; and Midwifery Today in the USA. Midwifery Today and the team surrounding Jan Tritten have established a unique experience of organising international conferences in countries as diverse as the USA, Jamaica, Japan, the UK, France and China.

The main theme of these conferences is always the rediscovery of authentic midwifery. Should the occasion arise, such groups have the capacity to raise questions in terms of civilisation, and not only in terms of individuals. For understandable reasons, this is not the case of the many consumer groups that participate in the natural childbirth movement. Consumer groups tend to promote choices in childbirth and to be the advocates of pregnant women. When a woman is pregnant, she tends to think of her own pregnancy and her own baby. As for publishers of magazines and books, they aim to reach the constantly renewed market of pregnant women.

This difficulty in enlarging the issue of childbirth might delay the advent of a new awareness. I can illustrate this difficulty by referring to my own experience. It is much easier to publish a book if the word 'birth' is included in the title. It is difficult to introduce the issue of birth indirectly. However those of my books I personally value are *Genese de l'homme écologique* (Genesis of an Ecological Human Being), *Primal Health* and *The Scientification of Love*. One of the most significant anecdotes is about *Genese de l'Homme écologique*, published in 1979, which was a book about the development of the respect for Mother Earth. Until the very last days the publisher tried to convince me to introduce the word 'birth' in the title. I did not take his advice into consideration for the French edition. Yet when the book was published in other languages, the word 'birth' miraculously appeared in the title. This is how *Genese de l'homme ecologique* became in German *Die Geburt des Menschen* (The Birth of Humans). I also noticed that when I am introduced in a group, it is commonplace to ignore the books that do not include 'birth' in the title.

In spite of the difficulties that are specific to childbirth, it was possible, until the end of the twentieth century, to concurrently look at industrialised farming and industrialised childbirth. The similarities between these two aspects of human activities were more visible than the differences. It is not so after entering the new millennium. The main difference is that a series of disasters have induced a new awareness in the case of farming. The history of childbirth has not yet reached the same stage. Which disaster are we waiting for?

8 WHICH DISASTER ARE WE WAITING FOR?

As a traveller I trust my personal rules of thumb. When staying in a city, I need to know how safe the place is. Can I walk in the streets after sunset? Since there is no easy access to reliable statistics of criminality, I just look at local birth statistics. My rule of thumb is that the rates of criminality are correlated with the rates of obstetrical intervention. This means, for example, that I'll be extremely cautious in places such as Sao Paulo, Mexico City, Rome or Athens, where the rates of caesarean sections are astronomical. On the other hand I'll be more relaxed on the streets of Tokyo, Stockholm or Amsterdam, where they have maintained comparatively low rates of obstetrical intervention. Cities such as London, Paris, Frankfurt or Sydney are between.

This rough common-sense approach is on the verge of being supported by scientifically established correlations. This will be a first step towards establishing a possible cause and effect relationship. If the media do not shun the data, the awareness-raising effect will be spectacular. I am just anticipating one possible scenario among hundreds. In reality nobody can predict what will trigger the 'eureka moment' where industrialised childbirth is concerned.

OTHER ANALOGIES

Until now disasters have been the most effective factors for consciousness-raising. Human beings had many new and powerful ways to play God during the twentieth century and to realise afterwards the drawbacks of their incapacity to think long term. Every day tons of carbon dioxide and other greenhouse gases have been pumped into the atmosphere as a result of burning fossil fuels. In the early 1970s I was already familiar with points of view commonly expressed in ecological publications. It was already clear for authors such as Murray Bookchin that humanity will be obliged in the near future to take fundamentally new directions based on renewable energy sources such as wind and solar power. A common-sense approach suggested that a continuous alteration of the atmosphere was bound to induce uncontrollable climatic changes. Since that time there have been countless warnings by authoritative scientific teams.

However we had to wait until the beginning of the new millennium to witness spectacular popular reactions, when climatic catastrophes became increasingly common. We are entering a phase in the history of mankind when the most powerful political leaders are about to realise that the health of the planet should prevail over many other considerations.

A USEFUL FORECASTING TOOL

In the current scientific context, we are in a position to forecast what sort of disaster will bring to light the dangers of industrialised childbirth. Everybody has easy access to our

Primal Health Research data bank (www.birthworks.org/ primalhealth). This data bank contains hundreds of references and abstracts of studies published in authoritative medical or scientific journals. All of them are about the long-term consequences of what happens during the 'primal period'. The primal period includes foetal life, the period surrounding birth and the year following birth. It is not easy to detect such studies because they do not fit into the current classifications. This is the main reason for the data bank.

From an overview of the bank it is immediately apparent that, in all fields of medicine, there have been studies detecting correlations between an adult disease and what happened when the mother was pregnant. It is even possible to conclude, through so many such studies, that our health is to a great extent shaped in the womb. But, because the most spectacular and invasive impact of industrialisation is on the very day when the baby is born, we must first trace studies establishing links with the birth itself.

THE DAY OF BIRTH

It is easy to detect such studies via listed key words such as birth complications, resuscitation, obstetric analgesia, obstetric medication, labour, labour induction, foetal distress in labour, cesarean delivery, asphyxiation, forceps, vacuum, cephalhaematoma … . By typing such key words we can detect a certain number of conditions that seem to be related to the period surrounding birth. It immediately becomes clear that looking at the long-term consequences of the manner in which we are born means going into the field of sociability, aggressiveness or, to put it another way, capacity to love. More precisely it appears that when researchers explore the background of people who have expressed some sort of *impaired capacity to love* – either love of oneself or love of others – they always detect risk factors at birth. 'Impaired capacity to love' is a very convenient term to underline the links between all these conditions. Furthermore when researchers find risk factors in the period surrounding birth, it is always about a very important issue specific to our time.

Juvenile violent criminality is undoubtedly topical. It can be regarded as an 'impaired capacity to love others'. It is not surprising that, according to a large and authoritative study by Adrian Raine, birth complications are among the risk factors for becoming a violent criminal at age 18.

There are many ways to express an impaired capacity to love oneself. Of course the most spectacular form of self-destructive behaviour is suicide and the most topical form of suicide is the suicide of teenagers. It is highly topical because it is a new phenomenon, unknown in other cultures. Today, in all industrialised countries, it is one of the main causes of death in adolescence. According to one of the most reliable evaluations that used data from the National Bureau of Statistics, the rate of suicide deaths among Australian males aged 15–24 years increased from 8.7 per 100,000 in 1964 to 30.9 per 100,000 in 1997.

One can conclude from these data that the risk of committing suicide depends upon the phase of the history of obstetrics when the teenager was born. It is significant that the only study of teenage suicide in our data bank detected risk factors on the day when the subject was born. Resuscitation at birth was one of the significant risk factors. There is also food for thought in the results of a series of studies by

Bertil Jacobson, from Sweden, who looked at the methods used to commit suicide. It appears that those who had a birth difficult from a mechanical point of view tend to use violent mechanical means if they commit suicide (jumping from a height, jumping in front of a train, shooting themselves, and so on). On the other hand suicides involving asphyxiation were closely associated with asphyxiation at birth.

There are more subtle forms of self-destructive behaviour, such as drug addiction, which is also highly topical. According to a series of studies in Sweden and in the USA, the risks of becoming drug addicted are increased among those whose mother had used certain pain-killers when she gave birth. As for anorexia nervosa, it can also be presented as an impaired capacity to love oneself. It is particularly common in our societies. A huge study, at the level of the whole Swedish female population, revealed the importance of risk factors at the very time of birth. The most significant risk factor – statistically speaking – is the fact of being born with a cephalhaematoma, that is a bloody collection inside one of the bones of the skull: it indicates that from a mechanical point of view the birth was difficult.

Autism can also be considered an impaired capacity to love. Autistic children and autistic adults do not socialise. When teenagers they cannot manage dating. When adults they do not have children. My interest in autism started in 1982, when I met Niko Tinbergen, who shared the Nobel prize with Konrad Lorenz and Karl Von Frisch. As an ethologist familiar with the observation of animal behaviour, he studied in particular the non-verbal behaviour of autistic children. As a field ethologist he studied the children in their home environment. Not only could he offer detailed descriptions of his observations, but at the same time he listed factors which predispose to autism or which can exaggerate the symptoms.

He found such factors evident in the period surrounding birth as: deep forceps delivery, birth under anaesthesia, resuscitation at birth and induction of labour. When I met him he was exploring possible links between difficulty in establishing eye-to-eye contact and the absence of eye-to-eye contact between mother and baby at birth. The work of Tinbergen (and his wife) represents the first attempt to explore autism from a primal health research perspective.

It is probably because I met Niko Tinbergen that I read with special attention, in 1991, a report by Ryoko Hattori, a psychiatrist from Kumamoto, Japan. Mrs Hattori evaluated the risks of becoming autistic according to the place of birth. She found that children born in a certain hospital were at increased risk. In that particular hospital the routine was to induce labour a week before the expected date of delivery and to use a complex mixture of drugs during labour.

BEFORE BIRTH

Industrialised childbirth also implies a certain style of prenatal care, constantly focusing on potential problems. Countless tests are routinely offered to all pregnant women, at different stages of their pregnancy. Simple physiological adaptive reactions are presented as diseases and named with bizarre terms. For example a transitory modification of the metabolism of carbohydrates is called 'gestational diabetes'. An increased blood volume, which is a good sign of placental activity, is misinterpreted as anaemia because the blood is

more diluted than usual, and the concentration of substances such as haemoglobin is therefore lower. It is obvious
that repeated prenatal consultations often have spectacular
negative effects on the emotional state of pregnant women,
planting seeds of doubt. I call that a Nocebo effect. That is
why, when exploring our data bank, we must also look at
studies evaluating the long-term consequences of altered
emotional states in pregnancy.

Several of these studies suggest that it is also in the fields
of sociability, aggressiveness or – to put it another way –
capacity to love, that the emotional states of the pregnant
woman may have long-term effects. The oldest of these
studies come from Finland. Two psychologists identified 167
children whose fathers had died before they were born. They
also identified 168 children whose fathers had died during
the children's first year of life. Then they followed all these
children through 35 years of medical records. All the
children grew up fatherless. Only those who lost their father
while in the womb were at increased risk of criminality,
alcoholism and mental disease.

Studies of children from unwanted pregnancies provide
similar conclusions. At the end of the 1950s a team from
Gothenburg, Sweden began a study to investigate from the
social-psychiatric viewpoint the lives of children who were
born after their mothers had applied for abortion but had the
application refused, and 240 people were first followed up
until the age of 21. Then the follow-up was extended to
completion of the thirty-fifth year. The main conclusion was
that the degree of sociability was lower in the group whose
mothers had unsuccessfully applied for an abortion. The
differences were still detectable at the age of 35.

The Prague study is based on a group of 220 subjects born

to mothers who, between 1961 and 1963, were refused an abortion both on initial request and subsequent appeal. The results of four waves of assessments were published. At age 30, 190 women were examined with pair matched control subjects. As in Sweden the degree of sociability was lower in the study group. The design, the objectives and the size of a Finnish study were different: 11,000 pregnant women were originally included in the study in 1966. In the sixth or seventh month of pregnancy mothers were asked whether the pregnancy was wanted, mis-timed but wanted, or un-wanted. The risk of later schizophrenia was significantly raised in the babies born to mothers in the unwanted group compared with the other groups.

Of course the alterations in the emotional states induced by industrialised prenatal care are not of the same nature and of the same magnitude as those which are mentioned in our data bank. However we can conclude that it is in the fields of behaviour and sociability that we should expect long-term consequences. Furthermore common sense suggests that it is arbitrary to distinguish correlations between what happened when the mother was pregnant and what happened when she gave birth. For example when a baby is made more fragile before being born through the stress hormones released by its mother, it is probable that the risks of foetal distress during labour are increased. Only the birth complications may be noticed and recorded. It does not mean that the beginning of the chain of events started on the day of birth.

Anyhow, when the 'eureka moment' occurs, the triggering factors will probably be in the field of behaviour, sociability, aggressiveness ... capacity to love.

GUESSING

While it seems easy to forecast which direction the events that might induce a new awareness will come from, the timing and the exact nature of such events are unpredictable. There have already been missed opportunities. For example when studies were published relating drug addiction to obstetric medication, one can imagine that some influential journalists would have picked up the data and would have claimed: 'Ah! Ah! Now we can understand why there has been such a need for addictive drugs among young Americans born in the age of twilight sleep.' Let us recall that twilight sleep included a shot of morphine.

There has been another recent example of missed opportunities. In France policemen (both 'policiers' and 'gendarmes') organised spectacular demonstrations because they cannot cope with the unprecedented number of crimes. The overall crime rate in France increased 7.69 per cent between 2000 and 2001, while the number of violent crimes increased 8.04 per cent. All French political parties agree that more policemen are urgently needed. Imagine that an influential journalist had jumped at the opportunity to look at the number of policemen in different countries in relation to how babies are born.

This would lead to look first at the statistics in at least two other comparable Western European countries: Holland and Italy. Holland is special, because 80 per cent of the midwives are independent and therefore powerful, while there is a comparatively small number of well-trained obstetricians, who are real experts in unusual or pathological situations. The Dutch statistics are unique, with about 30 per cent home births and an overall low rate of operative deliveries. Italy, on the other hand, is also special because it has the greatest number of obstetricians in Europe, compared with the population, one of the lowest number of midwives, and the highest rates of operative deliveries. France is in an intermediate situation regarding the number of obstetricians and the rates of operative deliveries. There are more midwives than in Italy, but they are much less powerful than the Dutch midwives. A typical French midwife is just a member of a sophisticated medical team in a large conventional department of obstetrics. The rates of interventions are much higher than in Holland, but lower than in Italy.

Such inquiries would inspire fruitful questions. How can we explain that in Holland, a country of 16 million people, they can manage with a total of 40,000 men and women employed with the police (2.5 per 1000 inhabitants)? At the same time, in France, they cannot manage with 220,000 policemen (policiers plus gendarmes) for a population of 61 million (3.6 per 1000). The discrepancies are still more spectacular when comparing Italy and Holland, particularly when looking at the official number of criminal cases a year (41 per 1000 in Italy versus 15 per 1000 in Holland).

Today one can just try to guess what the next opportunities will be.

My first guess is that alarming statistics will be widely published all over the world about the incidence of one of the countless aspects of impaired capacity to love. Some of them have never been studied from a primal health research perspective and are not mentioned in our data bank. At the very time when speculations will go apace, a well-designed large study will be published in a prestigious medical or scientific journal. This study will clearly demonstrate the strong links between this particular topical condition and one typical aspect of industrialised childbirth.

There are major difficulties in designing, conducting and publishing this family of research. The main one is that studies exploring correlations between an adult condition and obstetrical practices are at the limits of political correctness. This is the conclusion of my conversations with researchers such as Niko Tinbergen and Ryoko Hattori, who studied autism, Bertil Jacobson, who studied self-destructive behaviour in general and drug addiction in particular, Lee Salk, who studied the suicide of teenagers and Adrian Raine, who studied juvenile criminality. All of them had to overcome countless bureaucratic obstacles, including obstacles at the level of ethical committees.

I recently coined the term 'cul-de-sac epidemiology' when referring to such studies that often are not replicated, even by the original investigators. This framework includes research about topical issues. Despite publication in authoritative medical or scientific journals, the findings are shunned by the medical community and the media. I used this term to contrast with the term 'circular epidemiology', that has been used in order to condemn a tendency to constantly repeat the same studies, even when there is no doubt about the results. I came to the conclusion that when a study is politically correct, it leads to circular epidemiology. When it is not politically correct it leads to cul-de-sac epidemiology. Cul-de-sac epidemiology is a reason why a new awareness might be dangerously delayed.

Studies that are crucial to the future of mankind are not politically correct. How do we break such a vicious circle? Do we really need to wait for disasters? Would it be more appropriate to claim that we are unable to interpret disasters that are already developing?

Our predictions and guesses are mostly based on an overview of the Primal Health Research data bank. Other newly developed scientific disciplines also participate in this aspect of the current scientific revolution I call 'the scientification of love'. What can we learn from them?

9   THE SCIENTIFICATION OF LOVE

Love was traditionally the realm of poets, artists, novelists, philosophers and holy scriptures. At the dawn of the twenty-first century, it is studied from a variety of scientific perspectives. It is easy to miss the importance of the phenomenon because there is a multitude of specialised approaches to exploring the nature of Love.

NEW QUESTIONS AT THE DAWN OF
A NEW MILLENNIUM

Genuine scientific advances always inspire new questions. This is the case of the 'scientification of love'. During millennia all the possible means have been used to describe the different facets of love and to promote love. Countless philosophers pronounced on the nature of love. Paradoxically nobody was wondering how the capacity to love develops. Today we are prompted to ask this question because scientific data suggest answers. These data converge to give a great importance to early experiences, particularly to a short critical period immediately after birth. For similar reasons we suddenly think of wondering why all societies ritually disturb the first contact between mother and baby, for example by transmitting the belief that colostrum is tainted or harmful. We cannot help asking the question that way at a time when

we are learning from several new scientific perspectives that
the first hour following birth is probably critical in the
development of the capacity to love. One must keep in mind
that during many millennia the basic strategy for survival of
most human groups has been to dominate nature and to
dominate other human groups. There was therefore an
evolutionary advantage in developing the human potential
for aggression rather than the capacity to love. There was an
evolutionary advantage in disturbing the first contact
between mother and baby.

COMPLEMENTARY APPROACHES

Apart from primal health research, which provides crude
hard data, other disciplines are involved in the scientification
of love. They complement each other. When similar conclusions are simultaneously suggested through several approaches, they must be seriously taken into consideration.

A FIRST STEP

The first discipline which participated in the scientification
of love is, historically speaking, ethology. Ethologists observe
the behaviours of animals and human beings. Since the
emergence of their discipline, they traditionally have a
particular interest in mother–baby attachment. Whatever
the species of mammals they are studying, they always
confirm that there is a short yet crucial period immediately
after birth that will never be repeated. We must mention in
particular the work of Harlow, because he looked at mother

and baby monkeys, a species closely related to humans. Furthermore he followed up monkeys until adulthood and could establish correlations between different ways to disturb the first contact between mother and baby after birth and different alterations of sexual and maternal behaviour in adulthood.

HORMONES AND BEHAVIOUR

The scientification of love entered a new phase in 1968 when Terkel and Rosenblatt injected virgin rats with blood taken from mother rats just after they gave birth. The virgin rats behaved like mothers. Terkel and Rosenblatt had demonstrated that, immediately after birth, there are, in the blood of mother rats, hormones that can induce maternal love. This historical experiment was followed in the 1970s by a great number of studies exploring the behavioural effects of hormones such as oestrogens, progesterone and prolactine, whose levels are fluctuating in the period surrounding birth.

We had to wait until 1979 before learning that the hormone oxytocin has major behavioural effects. Before that time this hormone was known for its mechanical effects only. It stimulates uterine contractions for the birth of the baby and the delivery of the placenta. It stimulates the contractions of specialised breast cells for the ejection of the milk during lactation. Prange and Pedersen found that if oxytocin is injected directly into the brain of mammals, a maternal behaviour is induced. This experiment triggered a real explosion of research on the behavioural effects of this hormone. We can summarise the lesson we learnt from this generation of research by claiming that oxytocin is the typical altruistic hormone. It is the 'hormone of love'. Whichever facet of love we consider, oxytocin is involved. It was also in 1979 that we learnt about the maternal release of endorphins during labour and, in 1981, we learnt that the foetus also releases its own endorphins. It was already well known that all the morphine-like substances could induce states of dependency.

From that time it became possible to interpret the concept of critical period introduced by ethologists. We are now in a position to understand that all the different hormones released during labour are not yet eliminated during the hour following birth and that all of them have a specific role to play in the interaction between mother and baby. Just after birth mother and newborn are in a very complex specific hormonal balance.

We know that, in physiological conditions, the mother can reach a high peak of oxytocin – the hormone of love and also the hormone necessary for the delivery of the placenta. This peak of oxytocin is associated with a high level of prolactin – the motherhood hormone. The association oxytocin plus prolactin means love for babies. There are other circumstances when oxytocin is not associated with prolactin: it is another facet of love.

Immediately after birth, mother and baby are under the effects of a sort of morphine. When mother and baby are in close skin-to-skin contact, looking at each other, with their brains still impregnated with morphine-like hormones, it is the beginning of a dependency, that is the beginning of an attachment. It is impossible to offer a complete review of all the different hormonal agents involved in this early inter-action between mother and baby. It is too complex. Let us just underline that even hormones of the adrenaline family, which we commonly associate with aggressiveness, have a role to play immediately after birth. At the very time when the baby is born the mother is typically full of energy, with a tendency to be upright and to grasp something. This is the effect of a rush of adrenaline. That is why, immediately after a birth under physiological conditions, the mother still has a tendency to be upright, for example on her knees or seated on the floor. It is an advantage, among mammals in general, that the mother is alert, even aggressive, just after a birth. Aggressiveness is an aspect of maternal love. The baby itself needs to release its hormones of the adrenaline family during the powerful contractions of labour. One of the many effects of this 'noradrenaline' is that the baby is born with large pupils. This is a signal given to the mother, and it appears that the eye-to-eye contact is an important aspect of the mother–newborn relationship among humans.

This hormonal approach helps our understanding that sexuality is a whole. The same hormones are involved in the different episodes of sexual life such as intercourse, childbirth and lactation. Two groups of hormones are always involved – the altruistic hormone oxytocin, and the endorphins which can be considered to be our 'reward system'. Furthermore, similar scenarios are constantly reproduced. The final phase of each sexual event is always an 'ejection reflex': sperm ejection reflex, foetus ejection reflex, milk ejection reflex. This integrated vision of sexual life inspired by modern biological sciences suggests that when a cultural milieu interferes routinely with the birth process, it is in fact the entire sexuality – the entire capacity to love? – which is influenced.

EWES AND CIVILISATION

This reference to the cultural milieu offers an opportunity to underline that the scientification of love owes a lot to animal experiments. It is therefore relevant to clarify what we can learn from these experiments and also the limits of what we can learn. An example may be useful. It was found that when ewes give birth with epidural anaesthesia they do not take care of their lambs. We know that the effects of an epidural anaesthesia are much more complex among humans, and we know why. Human beings communicate with language. They create cultures. Their behaviour is less directly influenced by their hormonal balance. When a woman knows that she is expecting a baby, she can anticipate displaying some maternal behaviour. This does not mean that we cannot learn from non-human mammals. Animal experiments indicate the questions we should raise about ourselves.

Where human beings are concerned, the questions must include the terms 'civilisation' or 'culture'. If ewes do not take care of their babies after giving birth with epidural anaesthesia, this suggests that we should be wondering about the future of our civilisation if the birth process is routinely disturbed in that way.

Because we must think in terms of culture or civilisation rather than in terms of individuals, it is relevant to include in the scientification of love the ethnological approach. What can we learn by comparing cultures? Ethnology has established itself as a science by publishing databases. It is therefore possible to study the main characteristics of different cultures in relation to how babies are born.

The first conclusion of the ethnological approach is that all cultures disturb the physiological processes, particularly the first contact between mother and baby. I already mentioned that the most universal and intriguing way is simply to promote the belief that colostrum is tainted or harmful to the baby – even a substance to be expressed and discarded. Let us recall that, according to modern biological sciences, the colostrum available immediately after birth is precious. Let us also recall the newborn baby's ability to search for the nipple and to find it as early as the first hour following birth.

Several beliefs can be combined and reinforce each other. For example in some ethnic groups in Benin, West Africa, they transmit the belief that the mother must not look at the baby's eyes during the 24 hours following birth, so that the 'bad spirits' cannot enter the baby's body. The first contact between mother and baby can also be disturbed through rituals: rushing to cut the cord, bathing, rubbing, tight swaddling, foot binding, 'smoking' the baby, piercing the ears of the little girls, opening the doors in cold countries are examples of such rituals. The second conclusion of the ethnological approach is that the greater the need to develop aggression and the ability to destroy life, the more intrusive the rituals and cultural beliefs are in the period around birth.

This comparison between cultural milieux leads us, once more, to refer to the connection between birth and the mother–newborn relationship on the one hand, and capacity to love, sociability and aggressiveness on the other.

AN UNPRECEDENTED SITUATION

By combining all the perspectives involved in the scientification of love, it is easy to analyse the reasons why the history
of childbirth is at a real turning point. Although all societies
had in the past a tendency to take control of this event, the
situation is radically new at the dawn of the twenty-first
century.

Until recently a woman could not become a mother
without releasing a complex cocktail of love hormones.
Today, at the current phase of industrialised childbirth, most
women have babies without relying on this cocktail of
hormones. Some have a caesarean section which can be
decided and performed before the labour starts. Others block
the release of their natural hormones by relying on substitutes (usually a drip of synthetic oxytocin, plus an epidural
anaesthesia). Even those who eventually give birth without
any medication often receive a pharmacological agent for the
delivery of the placenta at a critical time in the mother–child
relationship. Let us stress that an injection of synthetic
oxytocin has no behavioural effect, because it does not cross
the blood–brain barrier. The questions inspired by such
widespread practices must be raised in terms of civilisation.

All these considerations must be framed in the context of
the twenty-first century. We are at a time when humanity
must invent radically new strategies for survival. Today we
are in the process of realising the limits of traditional
strategies. We must raise new questions such as: 'how do we
develop this form of love which is the respect for Mother
Earth?' In order to stop destroying the planet we need a sort
of unification of the planetary village. We need more than

ever the energies of Love. All the beliefs and rituals which challenge the maternal protective and aggressive instinct are losing their evolutionary advantages. This is the very time when the scientification of love is progressing. That is why this little-known aspect of the scientific revolution must be considered a landmark in the history of mankind.

10   BEES

From our quick review of the history of the twentieth century we have gained an understanding of a dangerous aspect of *Homo sapiens*. This aspect of human nature is an enormous discrepancy between the capacity to find solutions to old problems through clever, sophisticated and powerful techniques and, on the other hand, an inability to think long term and to anticipate the effects of a mass utilisation of emerging inventions. This unexplored human trait can be illustrated by countless examples.

UNDERPOLLINATED APPLES

The global shortage of bees in relation to the widespread use of powerful synthetic insecticides offers a significant example. Farmers have always tried to protect their crops from such enemies as insects, rodents and weeds. I had a personal experience of the use of traditional and time-consuming methods of potato field protection from a mass invasion by 'doryphores', or Colorado potato beetles. This was in a French village, in 1941, during the Second World War. The school children of the village had to spend hours every day gathering doryphores one by one by hand. Our role was considered so vital that we were now and then visited by the French authorities and also by the German soldiers. Having had such an experience I could easily understand the widespread and unambiguous enthusiasm of farmers when they were offered magic synthetic pesticides some years after the war, first in the USA and soon after in Europe. In 1947 more than 100,000,000 pounds of pesticides were already produced in the USA. An increased productivity was undoubtedly leading to much lower prices for fruits and vegetables.

At that time, as has already been noted, all these chemicals were also killing bees, bats, butterflies and birds. Such species play a key role in agriculture, transferring pollen from one seed to another. It has been well known for ages that bees, in particular, do much to fertilise the flowers of plants and trees, for some of the pollen from the male flowers gets brushed off on to the female flowers. These will then turn into seeds and fruit and produce new plants … these plants will in turn provide pollen and nectar for later generations of bees …. A great proportion of human diet comes, directly or indirectly, from crops that require, or are benefited by, insect pollination and bees are the chief pollinating insects. Not only have pollinator populations been hit hard by increased pesticide use, also much of their natural habitat, such as dead trees and old fence posts, has been destroyed to make more room for more farmland.

Today the global shortage of bees and other insects that pollinate plants is destroying crops around the world. These pollinators are so useful that bee-keeping has often had to be encouraged in places where the wild bees have been reduced in number. The point is to assess the economic ramifications if bees and other pollinators continue to disappear. The global food supply could be in serious jeopardy if the current tendencies are not reversed. Experts reckon that fruit and vegetables are bound to become more expensive, while their quality will change: an underpollinated apple usually means a smaller, less appealing apple.

The current situation was predictable and is not the consequence of a lack of knowledge. We were aware 50 years ago of the key role of insect pollination and it was already obvious that synthetic chemicals could not spare bees and other pollinators. Of course there has been a tiny number of 'eccentric' scientists or amateurs who started to discreetly raise non-politically correct questions. They were preaching in a desert because the variant of *Homo* I call *Homo* super-predator has a deep-rooted lack of interest in the future of our planet and a lack of compassion for the unconceived generations. This weak – or weakened – ecological instinct may be regarded as a form of impaired capacity to love. We must constantly go back to the basic question: how does the capacity to love develop?

THE SOIL POPULATION

We have too much to choose among all the possible examples of similar scenarios in the field of farming. Another devastating effect of industrialised farming is the deterioration of soil caused by the extensive use of nitrogen and pesticides, plus over-ploughing and straw burning. All soil must be considered as having a life of its own. In a fertile soil there is a vast multitude of creatures ranging in size from minute bacteria to large earthworms. There are many insects, some so small that a lens is needed to see them. Various kinds of symbiotic fungi help plants to take up micronutrients. All these living creatures work together as a harmonious whole, each having a special part to play. True stable humus is created by the soil population. John Soper used to say that the extent of the soil population can be grasped when one realises that, under a fertile pasture, their total weight is equal to that of the stock which can be carried. This is not recent knowledge: John Soper retired from the Colonial Agricultural Service in 1958 after 32 years in the tropics. Common sense could have anticipated the deterioration of the soils. Today experts reckon that in some places this deterioration is in fact an irreversible death. Some will underline that genetically modified crops can grow even in the desert.

We might chose, as another typical example of the blindness of the twentieth-century human being, the widespread use of antibiotics as growth promoters. They are given routinely to chickens, pigs and cattle. Tetracyclin, in particular, has been used primarily to promote animal growth on pig farms. Recently, researchers from the University of Illinois-Champaign showed that tetracyclin-resistant genes in the bacteria come from the hogs. Such findings are of paramount importance in terms of human health. The British Medical Association predicts that one of the biggest public health problems of the twenty-first century will be increased resistance to drugs. Sebastian Amyes, a bacteriologist with much experience in the field of bacterial resistance, gives us five years to discover new antibiotics or other measures. Otherwise he warns that 'we are going to slip further into an abyss of uncontrollable infection'. When I was a medical student, in the 1950s, I was already aware of the concept of antibiotic-resistant

microbes. Once more, common sense could have anticipated the current preoccupations.

VOLKSWAGENS

This blindness is not related uniquely to the particular
subject of farming. All sorts of human activities can
provide significant examples. The mass use of cars is a
symbol of the twentieth century. Replacing horses by
petrol engines to power four-wheeled vehicles was originally a fascinating invention, symptomatic of human clever-
ness. As early as the 1930s it became clear that the use of
cars was bound to become a mass phenomenon. The
emergence at that time of the Volkswagen – 'the people's
car' – and the designing of motorways, were highly
significant. However, even in the 1970s, the first warnings
about sending tons of carbon dioxide every day into the
atmosphere were ignored. Today about 800 millions cars
on the planet still work with fossil energy and contribute
hugely to the emission of greenhouse gases. Climatic
changes are becoming obvious, even if awareness is delayed
by group pressures and political leaders.

AN ADDITIONAL LESSON

The current lack of interest in the long-term consequences
of mass utilisation of new techniques leads to inescapable
questions about childbirth. The high rates of intervention
such as caesarean, induction of labour, augmentation of
labour, epidural anaesthesia and drug-induced delivery of

the placenta are not the mere effects of a deep-rooted misunderstanding of birth physiology. They are also a perfect illustration of the short-sightedness of technological man.

Medical circles justify the increasing rates of intervention by referring to scientific studies (particularly 'prospective randomised controlled studies') that evaluate through statistical methods the benefits and risks of obstetrical procedures. The point is that these studies, which are conducted in conventional obstetric departments, only consider short-term outcomes: the babies and the mothers are not followed up beyond the period surrounding birth. For example most obstetricians are now convinced that in the case of breech births at term it is better to routinely perform a planned caesarean. They also tend to dramatically reduce the number of women who can have a trial of labour after a previous caesarean section. They don't realise that for the first time in the history of humanity most women have babies without releasing a flow of hormones of love and that the future of our civilisation is at stake.

An additional lesson is inspired by the particular issue of industrialised childbirth. There is a contrast between the point of view representative of the male-dominated medical milieus and the point of view expressed by certain women. It is significant, for example, that the movement commonly called 'VBAC' (vaginal birth after caesarean section) was to a great extent initiated in the USA by Nancy Cohen, the author of *Silent Knife*, who herself had a home birth after a previous caesarean. Many doctors cannot understand why certain women do not want to have their labour induced, or why they try to give birth vaginally while their term baby is in a breech position, or why they are reluctant to receive a drug for the delivery of the placenta. Such a contrast suggests that the respect for the laws of nature might be partly a male–female relationship issue.

The respect for the laws of nature undoubtedly has physiological bases. Men, who release more testosterone, tend to be more aggressive and need to dominate both nature and other human beings. Women, on the other hand, go through a great variety of complex hormonal balances. When offering the breast to her baby, a woman is not in the same hormonal equilibrium as she was when giving birth or at the moment of the first contact with the newborn, or during intimate moments with her partner.

We must keep in mind that in traditional societies women have many babies and breastfeed each of them for several years. This implies that during a great part of their adult life they are above their basic level of the mother-hood hormone prolactin. The complex behavioural effects of prolactin have been well studied, particularly in Sweden. Prolactin tends to engender subordinate and submissive states of mind. In such states of mind the adaptability to the needs of the baby is increased. The laws of nature in general are then more easily accepted.

Given that the characteristics of a culture are shaped by the population's average hormonal balance, we should ask ourselves what is unique to our society at the beginning of the twenty-first century. One notable characteristic is that we have few children. Another is that the period of breastfeeding is brief, being completed in a matter of months. In most other societies, breastfeeding continues for years. In other words, high levels of prolactin are released for a very short period in the lives of modern women. Is the hormonal profile of women becoming more masculine?

How do we overcome this catch-22 situation?

11   FALLING ASLEEP AND FALLING IN LABOUR

TOWARDS A REDISCOVERY

Not only does this speedy review of the history of the twentieth century tell us about human nature, it has also practical implications where childbirth is concerned. We are given new reasons to disturb as little as possible the birth process and the first contact between mother and baby. We are given new reasons to improve our understanding of the factors that can make birth as easy as possible. In other words we need to rediscover the basic needs of labouring women. It has to be a real rediscovery, since all societies interfere with the physiological processes. We do not have any cultural model.

We must rely on the point of view of physiologists in order to understand these basic needs. Physiologists study what is cross-cultural and therefore universal. They help us to go back to the roots. They offer a sort of reference point from which we cannot deviate too much without taking the risk of uncontrollable negative side effects.

The first step must be to visualise a labouring woman with the eyes of a modern physiologist. This leads to *focus on the most active part of her body,* that is the gland secreting all the hormones involved in childbirth. These hormonal agents originate in old, primitive brain structures called the hypothalamus and the pituitary gland. In other words, if we visualise a labouring woman with the eyes of a modern physiologist, we visualise the deep, primitive part of her brain that is working hard and releasing a flow of hormones. Today we are also in a position to understand that when there are inhibitions – during the birth process or during any sort of sexual experience – such inhibitions originate in the 'new brain', the part of the brain which is highly developed among humans and which can be seen as the brain of the intellect. It is more appropriate to call it the new cortex or, rather, the neocortex.

When looking at the labouring woman with the eyes of a physiologist it is easy to interpret a phenomenon which is well known by certain mothers and by midwives who have the experience of undisturbed birth. It is the fact that when a woman is giving birth by herself, without any medication, there is a time when she has an obvious tendency to cut herself from our world, as if 'going to another planet'. She dares to do what she would never dare to do in her daily social life, for example screaming or swearing. She can find herself in the most unexpected postures, making the most unexpected noises. This means that she is reducing the control by her neocortex. This reduction of neocortical activity is the most important aspect of birth physiology from a practical point of view. It is the key to understanding that a labouring woman needs first to be protected from any sort of stimulation of her neocortex. What does it mean in practice?

AN ANALOGY

The best way to rediscover the basic needs of a labouring woman is to start from a comparison. Both falling asleep and 'falling in labour' are changes of states of consciousness. Both imply a reduced activity of the neocortex. The conditions necessary for the brain of the intellect to take a back seat are well understood when trying to sleep. They are forgotten where giving birth is concerned. That is why we need an analogy. Let us say that falling asleep is our 'source' and falling in labour is our 'target'. Our source is reliable because we have all learnt from experience how to protect ourselves from any sort of useless neocortical stimulation in order to fall asleep. We fall asleep at least once every 24 hours.

We all know that it is easier to fall asleep in a *silent place*. It is particularly difficult if somebody is talking, especially asking questions. When we feel sleepy at night, it is not helpful to answer questions about the maiden name of our mother or our fax number. We know that a last-minute conversation can postpone the time when we fall asleep. Language stimulates the neocortex, particularly rational language. When we communicate with language, we process what we perceive with specifically human brain structures. This point is more often than not ignored in the field of childbirth. In modern hospitals midwives or other members of staff ask labouring women questions because they must fill out forms. They must follow protocols designed by people who have not understood birth physiology. Women who are already 'on another planet' are asked, for example, about their haemoglobin concentration or about the time of their last meal. Many men do not hesitate to talk to their wife/partner when she is already in hard labour. There is a widespread misunderstanding of the effect of such language.

In the same way we all know that it is usually easier to fall asleep with a *dim light*, rather than with a bright light. We tend to switch off the lights at night. We invented blinds, shutters and curtains to protect our sleep. Practitioners who explore the activity of the brain with techniques such as electro-encephalography know how to stimulate the cortex of their patient. They switch on a light and they ask the patient to keep his (her) eyes open. Yet, in the age of industrialised childbirth – which is also the age of electric light – most women give birth under bright lights. Read textbooks for doctors or midwives. You will never find a chapter on this issue. This mere fact is a sign of a lack of understanding of birth physiology. When you look at a labouring woman from the perspective of a physiologist you anticipate that the intensity of the light is probably not neutral. It is at least a topic we should seriously discuss. Of course there are hospitals where they agree to dim the light when a woman is in labour. More often than not it is just to please a client who made a particular request rather than an initiative included in the hospital protocol in order to make births as easy as possible.

We also know how *difficult it is to fall asleep when we feel observed*. In other words *privacy* is a basic need. Imagine that a scientist wants to study with a video camera your positions while you are asleep. I doubt that you'll have a good night's sleep. The effect will probably be similar if you know that the rhythm of your heart is continuously recorded from sunset to dawn. There have been scientific studies suggesting that feeling observed is a situation which stimulates the neocortex. It is common sense. When we know that we are observed, we tend to correct our attitude. We feel different.

Such considerations suggest that we must contrast a midwife staying in front of a labouring woman and another one who is just sitting in a corner. They also suggest that, in a birthing place, we must be cautious with all devices that can be perceived as different ways to observe. It can be a camera introduced by the baby's father. It can be an electronic foetal monitor introduced by the doctors.

The need for privacy is an opportunity to refer again to the deep-rooted widespread misunderstanding of birth physiology. The tendency to ignore or to deny the need for privacy during labour is really cultural and not particular to one medical or non-medical milieu. Look at the many books about childbirth for the general public. They are often illustrated with pictures that transmit the wrong message: a woman in labour surrounded by two or three people watching her.

Many obstetricians were surprised when examining the results of a series of studies about the continuous recording of the rhythm of the baby's heart beats via electronic monitoring. All these studies confirmed that the electronic method has no other constant and significant effects on statistics than to increase the rates of caesarean sections, compared with intermittent auscultation, which means listening now and then. Many doctors could not anticipate such results. However it is obvious that if a labouring woman knows that her body functions are continuously monitored, there is a stimulation of her neocortex. This stimulation tends to make the labour longer, more difficult and therefore more dangerous, and more babies need to be rescued by caesarean section. It is noticeable that privacy during labour is not a specifically human need. All mammals have a specific strategy not to feel observed when giving birth. It is ironic that non-human mammals, whose neocortex is less developed than ours, know better than us what to do to put it at rest.

We also know how difficult it is to fall asleep when we feel threatened by any sort of danger. When we are aware of a possible danger we release hormones of the adrenaline family. Our neocortex is stimulated so that we are alert and attentive. We wake up in the middle of the night if it is getting cold, another situation associated with increased levels of adrenaline. It is the same when a woman is in labour. *She needs to feel secure.* Physiologists just help us to rediscover the basic needs in such circumstances. They cannot give recipes for feeling secure. However we can refer to a strategy women have used all over the world. They always had a tendency to give birth close to their mother, or close to somebody who can play the role of the mother, usually an experienced mother or grandmother in the community. This is the root of midwifery. A midwife is originally a mother-figure and our mother is the prototype of the person with whom we feel secure without feeling observed or judged. We cannot help thinking of children who need to feel the presence of their mummy at bedtime.

The analogy with falling asleep can help us in *rediscovering authentic midwifery.* In certain countries, particularly in Latin America, the midwives have almost completely disappeared. In North America there is a resurgence of midwifery after an eclipse. In other countries there are still many midwives, but the industrialisation of childbirth has dramatically altered their original role. Everywhere there is a deep-rooted misunderstanding of the very nature of mid-wifery that is equal to the misunderstanding of birth physiology. The need to rediscover authentic midwifery is obvious when analysing the misleading vocabulary commonly used, particularly in America, when referring to the birth attendant. Those who have understood the basic needs of labouring women would never use words such as 'coach' or 'coaching'. How can we 'coach' an involuntary process? 'Emotional support' is probably the most noxious term because it is constantly used. Like the little girl who needs to feel the presence of her mummy at bedtime, the woman in labour needs to feel secure without being observed. Nobody would say that the little girl needs a 'support person'. The word 'support' suggest an active role of the birth attendant. I was witness to an electronic conversation between Canadian doulas (a new term for a female birth attendant who is not officially a midwife) and midwives. 'Her partner is of little support and she doesn't have much family around her. There will be a primary midwife, secondary and apprentice there also. I was honored that the midwife recommended me as a doula, but more happy that this midwife looked at the mother's needs and is trying to help her cover all her bases and feel adequately cared for. She will have good labor support not just from one or two people, but from many ... .' I doubt that this woman, with so much 'support', gave birth easily.

The collective misunderstanding of human birth is infectious and has contaminated many different specialised circles. For example the theoreticians who want to interpret the difficulties of childbirth in our species only think of mechanical problems in relation to the size and the shape of the maternal pelvis. They seem unable to turn their attention above the waist. They never think of this flow of hormones that must be released by the primitive brain. They never refer to the powerful inhibitions that originate in the neocortex. Today even some of those who take care of non-human mammals share this misunderstanding. There are many stories of catastrophic births among endangered species in zoos. Journalists and photographers are introduced without any precaution. In a zoo in the United States, an elephant needed a dramatic caesarean after the labouring phase had been observed by several people (including the trainer who was supposed to be the 'support' person).

THE LIMITS OF AN ANALOGY

The analogy with falling asleep is instrumental in formulating in a simple and concise way the conditions needed for effective labour to establish itself properly: being protected from useless words, being sheltered from bright lights, being in an atmosphere of privacy, feeling comfortable in terms of temperature and feeling secure.

We must separate from this analogy at the time of the 'foetus ejection reflex', that is the last two or three irresistible contractions before the birth of the baby. Once more we refer to the physiological reference point. This reflex is inhibited if another person takes on the role of coach, guide, helper, support person or observer. At that stage it is more relevant to underline the similarities with orgasm. I heard at least a dozen women who spontaneously used the term 'orgasm' to refer to the birth of their baby. The flows of hormones are comparable.

We must also separate from this analogy during the

so-called third stage, which is between the birth of the baby and the birth of the placenta. This phase of interaction between mother and baby has been highly disturbed by human societies. Yet it is a critical time for the survival of the mother, because a difficult separation and delivery of the placenta can lead to a life threatening haemorrhage. It is also a critical time for mother–child attachment. From a hormonal point of view this phase is characterised by the capacity the mother has to suddenly release a high peak of oxytocin, the hormone necessary to contract the uterus and also the typical hormone of love. When women bleed due to difficult delivery of the placenta, it is because they have not reached a high enough peak of oxytocin at the right time. At that time bleeding is the consequence of an inappropriate environment. This is why it is so important to clarify the conditions for an effective release of oxytocin.

The first condition for a safe delivery of the placenta is that the birthing place is warm enough. We must keep in mind that adrenaline – a hormone we release when it is not warm enough – is antagonistic of oxytocin. After the birth of the baby women rarely complain that it is too hot. If they are shivering, it is because it is not warm enough. When pregnant women ask me what to prepare for a home birth, I only mention a transportable heater that can be plugged in at any place and at any time. It can be used to warm up blankets and towels to cover the bodies of mother and baby.

The second condition is that the mother is not distracted at all and has nothing else to do than to look at the baby's eyes while feeling the contact with the baby's skin. This is the difficult point, because as soon as a baby is born, there is usually an irrational need for activity around. The mother can be distracted by somebody who is talking, or watching her, or switching on a light, or by an untimely phone call, or by somebody who wants to cut the cord, and so on. One of the reasons why cutting the cord before the delivery of the placenta is dangerous is that it is a distraction for the mother and it interferes with the mother–baby interaction.

Because these two conditions for a safe delivery of the placenta are ignored, thousands of mothers die every year, particularly in third-world countries. For similar reasons obstetricians are right to routinely inject, as soon as the baby is born, a drug replacing the natural maternal oxytocin. This is a way to compensate for the effects of an inappropriate environment, which is itself the consequence of a lack of interest in birth physiology. In terms of civilisation, injecting routinely a substitute for the love hormone at such a critical time is *one of the most threatening aspects of industrialised childbirth.*

12 IS THE PARTICIPATION OF THE FATHER AT BIRTH DANGEROUS?

The participation of the father at birth is undoubtedly an aspect of industrialised childbirth. A century ago, when most babies were born at home, such a question would have been deemed irrelevant. At that time, everybody knew that childbirth is 'women's business'. The husband was given a practical task, such as spending hours boiling water, but he was not involved in the birth itself.

Today, at the height of industrialised childbirth, the same question is still deemed irrelevant, even stupid. At the dawn of the twenty-first century, everybody knows about the importance of the active role of the father in the 'birth of a family'. Most women cannot even imagine giving birth without the participation of their partner. We have heard countless wonderful stories of 'couples giving birth'. Fathers are welcome in conventional delivery rooms.

In order to interpret such sudden and radical changes in point of view and behaviour, one must put them into their historical context. It is essential to recall that this intriguing phenomenon began unexpectedly in most industrialised countries in the 1960s. Then, a new generation of women felt the need to be assisted by the baby's father when giving birth.

They started to express this new demand at the very time when births were more and more concentrated in larger and larger hospitals. Birth in huge maternity hospitals has been an important step in the history of industrialised childbirth. This was also the time when the midwife became one of the members of a large medical team (in the countries where she had not completely disappeared). It is clear that the participation of the father was as an adaptation to unprecedented situations. It had not happened before in the history of mankind that women had to give birth in large hospitals among strangers; as for midwives, they had always been independent.

Those who have been active witnesses of such behavioural upheavals remember how quickly theoreticians established new doctrines. For example, I heard in around 1970 that the participation of the father would strengthen ties inside the couples and that we should expect a decrease in the rate of divorces and separations. I also heard that the presence of the father, as a familiar person, should make the birth easier and that we should expect a decrease in the rate of caesareans.

In order to prepare for a new era in childbirth, we must reconsider the behaviour and theories that are historically associated with industrialised childbirth. We must take an inventory of the questions we must raise. Where the participation of the father at birth is concerned, we must raise at least three questions:

First question: does the participation of the father aid or hinder the birth?

Those who are old enough to remember what a birth can be like when there is nobody around other than an experienced, motherly and low-profile midwife are inclined to formulate the question that way. Our objective is not to provide answers but to analyse the many reasons why it is such a complex issue.

There are many kinds of couples according to the duration of cohabitation, the degree of intimacy, and so forth. There are many kinds of men; some can keep a low profile while their partner is in labour; others tend to behave like observers, or like guides, whereas others are much more like protectors. At the very time when the labouring woman needs to reduce the activity of her intellect (of her neocortex) and 'to go to another planet', many men cannot stop being rational. Some look brave, but their release of high levels of adrenaline is contagious.

The double language of human beings appears as the main reason why the complexity of such issues is underestimated. There is a frequent contradiction between the verbal language and the body language of pregnant women. With words, most modern women are adamant that they need the participation of the baby's father while they give birth; but on the day of the birth the same women can express exactly the opposite in a non-verbal way. I remember a certain number of births that were going on slowly up to the time when the father was unexpectedly obliged to go out (for example to buy something urgently before the store is closed). As soon as the man left, the labouring woman started to shout out, she went to the bathroom and the baby was born after a short series of powerful and irresistible contractions (what I call a 'foetus ejection reflex').

When raising such a question one must also take into account the particularities of the different stages of labour. Some women feel inhibited at this phase of labour when they empty their rectum ... an opportunity to stress that the kind of intimacy a woman can have with her sexual partner is not of the same nature as the feeling of privacy she can have with her mother. It is often between the birth of the baby and the delivery of the placenta that many men have a sudden need for activity, at the very time when the mother should have nothing else to do than to look at her baby's eyes and to feel the contact with her baby's skin in a warm place. Let us repeat that at this time any distraction tends to inhibit the release of oxytocin and therefore interferes with the delivery of the placenta.

Second question: can the participation of the father at birth influence the sexual life of the couple afterwards?

Through such a question we introduce the complex issue of sexual attraction. Sexual attraction is mysterious. Mystery has a role to play in inducing and cultivating sexual attraction. Once there were mother goddesses. At that time childbirth was enigmatic among the world of men. I have had the opportunity in the past to talk about the birth of their baby with women who were themselves born at the end of the nineteenth century. They could not imagine being watched by their husband when giving birth: 'and what about our sexual life afterwards?' was their most common reaction.

Today I am amazed by the great number of couples who separate some years after a wonderful birth according to the modern criteria. They remain good friends but they are not sexual partners any longer. It is as if the birth of the baby had reinforced their comradeship while sexual attraction was fading away.

Third question: can all men cope with the strong emotional reactions they may have while participating in the birth?

In the age of industrialised childbirth, at a time when women can watch TV in the delivery room, it is uncommon to raise the question that way. During the days following an industrialised birth, nobody is wondering about the well being of the father. When visiting a family two or three days after a home birth, I almost always found a happy and active mother taking care of her baby. I had a surprise when asking about the father. More often than not I heard that the father was in bed, because he had a tummy ache, or a back ache, or flu, or a tooth ache, or simply because he was 'drained', as a mother told me. When referring to my experience of home birth, I am tempted to claim that male postnatal depression is common in a certain context, although it is not recognised as such.

The concept of male postnatal depression is a reminder that many cultures have rituals whose effects are to channel the emotional reactions of the father. All these rituals belong to the framework of the 'couvade' (anthropologists use this term that originally means, in French, 'hatching'). These rituals, whatever the local particularities, make the father busy while his wife is giving birth. The last example of couvade was the man spending long hours boiling water. I cannot help thinking of the case of young modern men who spend a long time assembling a rented transportable birthing pool: finally the baby is born before the pool is ready. Is it a revival of the couvade?

When our societies have reached a sufficient degree of awareness regarding industrialised childbirth, the routine participation of the father at birth will become a central theme of discussion.

13   HOW DANGEROUS IS

A CAMERA?

The frequent use of cameras is another aspect of industrialised childbirth. Photography is not new. It was invented in 1839. However, until the height of industrialised childbirth, as long as the birth of a baby was still women's business, nobody could even imagine taking a picture of a baby entering the world. The first picture was often made some hours after birth.

The first motivation for introducing cameras was to put forward pictures that showed alternatives to childbirth with the mother on a table, under bright lights, with her legs on stirrups, surrounded by several white-coated people. There was a time when it was imperative to do away with many of the old mental images associated with words like delivery and birth. In our hospital we ourselves have indeed played a great part in the spread of the epidemic of photos and films, even before the age of videos. But we were certainly aware of the need for privacy.

In participating in reports with photographs or TV programmes, we were very anxious to introduce the camera only at the very last moment, just before birth, at the point of no return when there is no risk of stopping the progress of delivery. In the context of a hospital with several births a day, it was possible to improvise and to introduce the camera only in a small number of highly selected cases, when the woman was really 'on another planet'. We always avoided making pictures during the first stage of labour, and we were very cautious before the delivery of the placenta. I remember one woman who delivered her baby right in front of a big German TV camera and said some minutes afterward: 'It was wonderful! What a pity there was nobody to take a picture!'

Now it is commonplace to use a camera, particularly a video camera, during a birth at home or in a birthing centre, without realising how invasive it can be. There have been countless TV programmes about home birth. It seems easy to find mothers-to-be who accept in advance the presence of a TV crew. It is significant that when there are complications in such circumstances, nobody thinks of relating the complications to the presence of a camera. I might report many meaningful anecdotes.

There is the story of a North American midwife who was sued after the death of a baby born at home. It was a breech presentation. It was easy afterwards to criticise the attitude of this poor defenceless midwife because the whole period surrounding birth had been recorded by a video camera. I just remember from the e-mail discussion of this case that nobody commented on the dangers of filming a birth. I have personal experience of breech births at home. I would never agree to participate in such a home birth if the mother were aware of the presence of a camera.

The current epidemic of photos and videos is first and foremost one of the symptoms of a widespread cultural misunderstanding of birth physiology. Today the priority is to rediscover the need for privacy. We must learn to eliminate all the onlookers and all their different ways of observing. That is why, in this book, there are no photos of birth. The presence of a camera is incompatible with a biodynamic attitude to childbirth.

14   TOWARDS A BIODYNAMIC
ATTITUDE TO CHILDBIRTH

WHAT DOES IT MEAN TO HAVE A BIODYNAMIC ATTITUDE?

Without waiting for widespread new awareness, we must start preparing for the post-industrialised age of childbirth. This implies radically new attitudes. New attitudes will be associated with a renewed vocabulary. This is a reason to prolong our analogy between farming and childbirth. I cannot find a better term than 'biodynamic' to describe the attitudes that should characterise post-industrialised childbirth. The term 'natural childbirth' is obsolete. It can only be used in retrospect, when a woman has given birth without any drug and without any intervention. We do not need to qualify an outcome. We need to qualify an attitude.

A biodynamic attitude is based on a good understanding of the physiological processes. The point is to take advantage of the whole physiological potential of mother and baby. It is the opposite of culturally controlled childbirth, which is in fact medically controlled childbirth in our society. The contrast between 'medically controlled childbirth' and a biodynamic attitude can be explained in terms of disparity between basic preoccupations, and also by comparing different ways of facing difficulties.

DIFFERENT PREOCCUPATIONS

Let us take a concrete example. Let's imagine a woman is now entering the phase of established labour. In the case of a medically assisted birth the main preoccupation will be to use the most effective methods for monitoring the foetus in order to be constantly in the best possible situation to rescue a baby in danger. The priority is to control what is happening. Such an attitude led to the concept of electronic monitoring, which became a symbol of industrialised child-birth. It is painful to get rid of symbols. That is why electronic foetal monitoring is still in use in certain hospitals, in spite of the many studies demonstrating that, compared with intermittent auscultation of the baby's heart-beats, it has no other constant and significant effects on statistics than to increase the rate of caesareans.

Let us imagine now the same phase of labour in a biodynamic atmosphere. The first preoccupation will be to reduce as much as possible the risk of foetal distress during labour. The point will be to make the birth as easy as possible. The mother-to-be will have complete privacy and will not feel guided or observed. She will feel free to be noisy and to be in the most unexpected postures. There will be a high probability that she will find herself on her hands and knees, or in any sort of bending forward posture. The first immediate effect will be to alleviate the pain, particularly if it is back pain. The most important mechanical effect will be the absence of compression of the big vessels that run along the spine by the heavy uterus: the most common factor for foetal distress during labour will be eliminated. At the same time the rotation of the baby's head will be facilitated. Let us add that when a labouring woman is on her hands and knees, as if praying, she can more easily cut herself off from the rest of the world and 'go to another planet'. In other words she can more easily reduce the activity of her intellect and reach the right hormonal balance. An experienced midwife will keep a low profile and, without being invasive, will find the right moments to listen to the baby's heart with a 'pocket Doppler'. For example she can take advantage of the time when the woman is walking towards the toilet. Whatever the posture of the mother, she can use (in appropriate circumstances and with moderation) this marvellous and cheap piece of technology, even though she knows that the risks of foetal distress are almost completely eliminated in such a context. The need for drugs will be dramatically reduced, and therefore another reason for foetal distress.

COPING WITH DIFFICULTIES

Let us take the example of a labour that is long, difficult and painful. This suggests that the woman in labour probably has difficulty in releasing the hormones involved in the birth process, particularly the pituitary oxytocin (necessary to contract the uterus) and the natural pain killers usually called endorphins. This is a common situation in the case of a medically controlled labour. The tendency will be to immediately replace the necessary hormones by pharmacological substitutes. A drip of synthetic oxytocin will replace the pituitary hormone, while an epidural anesthesia will be a substitute for endorphins.

Let us imagine now a similar situation in spite of a biodynamic attitude. The primary question inspired by a physiological perspective will be: can this woman reduce her levels of hormones of the adrenaline family? If the question is raised in such a simple way, it is possible that an appropriate response will immediately appear. Perhaps the room is too cold and it becomes urgent to heat it up. Or the mother-to-be cannot reach the phase of established labour because she is slightly hungry. Hunger is on a par with increased levels of adrenaline: a little snack will work wonders. Or, more often than not, there is somebody around who is releasing a certain amount of adrenaline. It is usually either the baby's father or the doctor, or both. An increased level of adrenaline is highly contagious. Complete privacy in semi-darkness with the proximity of a discreet and motherly midwife might be the best way to evaluate the real physiological potential of the labouring woman. It is also by raising the question that way that the use of a birthing pool, for example, can be suggested. Immersion in water at the temperature of the body is undoubtedly a way to reduce the level of hormones of the adrenaline family. Once I immersed myself in a birthing pool. I fell asleep. This was a reliable sign of a low level of adrenaline.

Midwives of the post-industrialised age will be familiar with the effects of water immersion. They will know how to take advantage of the birthing pool when the labour is difficult, although the basic needs of the labouring woman seem to be met. They will have digested a small number of simple rules. They will first give great importance to the time when the woman is anticipating the bath, that is the time when she can hear the noise of water and can see the beautiful blue water filling the pool: then many inhibitions can already be released. They will have understood how immersion in water at the temperature of the body can make uterine contractions more effective during a limited period of time – that is in the region of an hour and a half. The two main recommendations about the proper use of birthing pools are based on this simple fact.

The first recommendation is to help women to be patient enough to postpone the bath, ideally until a certain degree of dilation of the cervix. If water immersion starts when the cervix is about five centimetres dilated, it is almost a guarantee that the need for drugs and intervention is eliminated: complete dilation will probably be reached within an hour and a half, even in the case of a first baby. Experienced midwives are good at helping women to be patient enough. For example, while waiting for the bath to reach the right temperature (never above body temperature) the midwife can suggest a shower.

There are several reasons why a shower can be surprisingly effective. The main one is that while under the shower the woman in labour is more often than not by herself in a small space, that is in a situation of complete privacy. Also the noise of water can exert its magic power to release inhibitions. Furthermore the jet can be directed towards the nipples to stimulate the release of oxytocin, or towards the back to relieve back pain.

The second recommendation is to avoid planning a birth under water. The baby can be born under water when there are suddenly irresistible powerful contractions and the mother does not feel like getting out of the pool: it should not be the objective. The objective is to reduce the need for drugs. Often women need to get out of the pool for the very last contractions, at a phase when paradoxically a short rush of adrenaline can help. Women who are prisoners of the project of giving birth under water may be tempted to stay too long in the bath, so that finally the baby is born when the contractions are already weaker. Then the contractions will be still less effective for the delivery of the placenta.

The contrast between medically controlled childbirth and a biodynamic attitude is best illustrated by the way the birthing pool is used. In the case of a biodynamic attitude the tendency is to keep it at the disposal of labouring women and to offer it as a way to replace drugs, should the labour be long, difficult and very painful. The birthing pool can be used for a fast 'aquatic trial of labour', that is to say in order to decide if a baby may be born by the vaginal route. When there is no spectacular progress of the dilation of the cervix after an hour spent in the bath during hard labour, there is no reason for procrastination. It means that there is a serious problem and that a caesarean section will be needed. In the case of medically controlled childbirth the tendency is to multiply the contraindications for the use of the birthing pool, so that finally only the low risk women who don't really need it are 'authorised' to spend some time in water.

This simplified review of preoccupations and strategies associated with a biodynamic attitude indicates the direction of the shift that should follow the industrialised age of childbirth. How should we prepare for the shift?

15 THE FUTURE OF THE MIDWIFERY–OBSTETRICS RELATIONSHIP

THE NEED TO BE RADICAL

Any shift towards a biodynamic attitude needs to be strictly speaking radical, that is tackling the problem at its root. At the root of the problem is the medical control of childbirth, which is a modern variant of cultural control. This medical control is a corruption of the role of medicine. The role of medicine in general – and obstetrics in particular – is originally limited to the treatment of pathological or abnormal situations. It does not include the control of physiological processes.

During the twentieth century pregnancy and childbirth became for the first time the domain of medicine. This is how a wonderful rescue operation – namely the caesarean – became one of the most common ways to be born. This is how a drug useful to treat anomalies in the progress of labour – synthetic oxytocin – became the basis of 'active management of labour'. This is how an effective way to treat a pathological pain – epidural anaesthesia – became compatible with the term 'normal birth'. The medical control of childbirth has reinforced a widespread inability to think long term and to think in terms of civilisation. It has stifled the voices of women who were viscerally attracted towards alternatives to industrialised childbirth. It explains the gap between the demands expressed by certain pregnant women and the point of view of a great number of medically trained professionals: those who cannot understand why there are still women who want to go through the pain and stress of labour in the age of elective caesarean on demand, epidural anaesthesia and oxytocin drips. Today any shift towards a biodynamic attitude leads first to a total reconsideration of the role of obstetrics and the reason for obstetrics.

TOWARDS AUTHENTIC OBSTETRICS

We just need to look at some statistics to realise why obstetricians of the industrialised era cannot be reliable experts in unusual, strange or pathological situations. Let us take the United States as our example, since it has constantly preceded other countries in the trend towards industrialised childbirth. In the US, the number of obstetricians is in the region of 36,000, for a number of births a year that is in the region of 3,600,000. This implies that a typical obstetrician is in charge of about 100 births a year.

Most modern obstetricians are therefore primary care givers rather than doctors specialised in pathological or unusual situations. They have a dangerous lack of experience. For example a typical American obstetrician has the experience of about one twin birth a year. He (she) needs five years of practice to be confronted with one shoulder dystocia (when the baby is stuck at the level of the shoulders). He (she) needs ten years of practice to see one real placenta praevia (the baby cannot get out because the placenta is in the way) and a whole career to see one real eclampsia. On the day when she (he) has to do a caesarean for a transverse presentation, she must adapt her technique after referring to her text books, because it is a rare situation she probably has never met. In the maternity unit of a French hospital, I was in charge of about 1000 births a year, with six midwives. I had the feeling that it was the right number to maintain sufficient experience.

The prerequisite for the replacement of medically controlled childbirth by a biodynamic attitude is a dramatic reduction in the number of obstetricians. The highly trained experts of the future will not have the time to control every birth. They will be at the service of women and midwives. They will appear on demand.

LOOKING AT NUMBERS

A dramatic reduction in the number of obstetricians must undoubtedly be balanced by an appropriate increase in the number of midwives. This implies that after the phase of awareness there will be a phase of transition. Because it is to a great extent a question of numbers, a different proportion of midwives to obstetricians cannot be reached overnight. The period of transition might cover several decades.

The transition should be easier and faster in countries where there are already a comparatively great number of midwives and a moderate number of obstetricians. It is noticeable that such countries are those with the best possible statistics. For example, in Sweden, there are about 6000 midwives for a population of about 9,000,000 (compared with 5000 certified nurse midwives for the whole of the USA!) and a comparatively small number of obstetricians. Sweden has the best birth outcomes in the Western world, with a moderate rate of caesareans that remained pretty stable, around 11 per cent, for almost 20 years. In Holland, 80 per cent of the midwives are independent. When a woman is pregnant her reflex is more often than not to visit a midwife. The role of the midwife is to decide, during pregnancy and birth, if she needs the advice or the service of a doctor. Yet she is not answerable to any doctor. During labour the mother and the midwife can decide to stay at home and the rate of home births is around 30 per cent (it is below 2 per cent in all other industrialised countries). Less than 5 per cent of Dutch women presently need an epidural anaesthesia during labour and the rate of caesareans, in the region of 10 per cent, is the lowest in Western Europe.

In Japan, a country with many midwives, they never developed huge maternity hospitals with thousands of births a year, one of the main features of industrialised childbirth. The average number of births a year in a maternity hospital is in the region of 500. Japan has the lowest 'perinatal mortality rate' (number of babies who spent more than six months in the womb and died before the age of a week) in the world, with a moderate rate of caesareans and a rate of epidurals still lower than in Holland.

These comparisons between countries inspire many comments and questions. They clearly indicate that, in general, births are easier in countries where the doctor is usually kept away. Where childbirth is concerned there is such a disparity between countries with equivalent standards of living that inescapable questions will probably be raised in the near future. Researchers will have to look at the evolution of cultural characteristics in relation to how babies are born. One can already wonder why, for example, the streets of Amsterdam are safer than the streets of Paris, and why Holland has the lowest rates of abortion, imprisonment and teenage pregnancy in the West, with comparatively low rates of drug addiction in spite of the open sale of marijuana and hashish.

Such comparisons should also help in realising that the gap between countries will tend to become deeper, should drastic remedies be dangerously delayed. In countries where there is a small number of midwives and therefore a well-established medical control of childbirth, many midwives were born and gave birth in an industrialised environment. They have no personal, visceral experience of childbirth in physiological conditions. This leads to questions about the nature of authentic midwifery.

TOWARDS AUTHENTIC MIDWIFERY

The shift towards authentic midwifery is more than a matter of numbers. It implies a good understanding of the reason for midwifery. We must constantly go back to the questions inspired by physiological considerations: how can a woman in labour feel secure without feeling observed or judged? The presence of a mother can reconcile such needs. In the age of medically controlled childbirth, the midwife is perceived as an obstetric nurse, that is a member of a medical team. An authentic midwife should be perceived first as a mother figure. This leads to a central and inescapable question at the dawn of the post-industrialised age: how to select women who will enter the midwifery schools?

The policy that will transform the way babies are born is inspired by a simple observation that fits physiological considerations: when a woman gave birth vaginally to her own baby(ies) without any medication, this is a guarantee that her presence at a birth will not hinder the progress of labour. The crucial aspect of the drastic remedies we must resort to is the mode of selection of student midwives. We must radically reconsider the criteria currently in use. The prerequisite for entering a midwifery school should be a personal experience of unmedicated birth. Let us recall that in most traditional societies a midwife was a mother or a grandmother who had many children. Women who had many children were usually those who had easy births. Such a programme, easy to summarise in one sentence, is bound to face many difficulties and will have to overcome predictable obstacles.

The first major obstacle will be the usual reaction to the idea that an authentic midwife should have the experience of giving birth 'normally'. People immediately react by claiming that they know wonderful midwives who are not mothers. They are right. I know many of them and I have myself practised as a home birth midwife although, for obvious reasons, I'll never be a mother. The point is that we are preparing for the future and we are thinking in terms of civilisation. By selecting women who had a positive experience of the birth of their own baby(ies), we offer a guarantee that cannot be provided by any other mode of selection. Those who are in charge of the selection of midwifery students must overcome their personal inner conflicts. They must learn to see far into the future. They must also surmount many previous doctrines and received ideas. I heard several European midwifery teachers – including in

Holland – claiming that they prefer to accept young students who are not in charge of a family, and who are quite malleable before having a personal experience of life. They claim that the practice of midwifery is incompatible with motherhood. All these contradictions will gradually vanish when there is a much greater number of midwives, so that midwifery will usually become a part-time occupation. It is true that young ladies who have a limited and uniform background can more easily become proud of being highly specialised technicians. An authentic midwife is supposed to be first a 'wise woman' (sage-femme). Being a wise woman is the opposite of being a narrow technician.

After decades of industrialised childbirth, there will be another obstacle to adopting these radically new criteria of selection. We must realise that in many countries the number of women who had a positive experience of unmedicated vaginal birth is already insignificant. These are precisely the countries where there is an urgent need to develop many midwifery schools and to detect many potentially authentic midwives. In order to break the vicious circle, a policy will be needed that would insistently encourage the rare women who gave birth by themselves to become midwives, at least during a certain phase of their life.

These considerations about the future of the midwifery–obstetrics relationship will overturn the order of the main preoccupations humanity will have to face in the near future. It is not usual to give so much importance to the selection of midwifery students. It is not usual to claim that those who are in charge of choosing the midwives of the future probably have greater responsibilities than the best-known political leaders, where the future of our civilisation is concerned.

16   HAVING A BABY BEFORE 2032

Nobody knows when a mass demand for post-industrialised childbirth might emerge. In the age of powerful media it can be at any time. Yet the shift towards a biodynamic attitude to childbirth – that implies a shift towards authentic midwifery and authentic obstetrics – cannot be completed overnight. One might claim that in general the narrow-minded highly specialised technical man cannot rapidly digest the concept of a biodynamic attitude.

THE SEEDS FOR A WIDESPREAD AWARENESS

Meanwhile we must think of the young ladies who are now at the dawn of their reproductive life and who are therefore destined to have their babies within 30 years or so. Some of them are endowed with a deep-rooted belief in the importance of the way a baby is born. They precede the collective awareness we are waiting for. They cannot wait for major changes in the midwifery–obstetrics relationship. They must adapt to a period of transition.

It is noticeable that everywhere there are cores of people who are trying to challenge the current conventional attitudes. These people are the seeds for a widespread awareness. They are ready for the predictable growth spurt of an emerging movement. They are often instrumental in helping women to find the best local resources and options. Although the industrialisation of childbirth is a global phenomenon, there are such differences between countries that strategies must be adapted to geographical, political and historical particularities.

DIFFERENT WAYS TO PREPARE

There are different ways to prepare for the post-industrialised era of childbirth. One of them is to transform hospital maternity units into home-like places. This is what we tried to do in a French state hospital in the 1970s and 1980s. In the German-speaking countries of continental Europe there has been, since the beginning of the 1990s, the sudden emergence of dozens of free-standing birthing centres. Similar birthing centres appeared in other European countries, in Australia and in the USA, where the phenomenon started as early as in the 1970s. The concept of birthing centre is a way to go back to the roots. It is older than the concept of home birth. In traditional societies women were not giving birth at home. They were usually giving birth in special huts, where often they could also stay during their periods. In certain cultures the place of birth was the cattle house or the place where women could bathe. In Scandinavian countries some hospitals now have an ABC (alternative birthing centre) clinic, that is a place where you can forget that you are in a hospital.

Adapting home birth to the context of the twenty-first century might become another way to prepare for the post-industrialised era. Today a great number of women live in an urbanised environment, and therefore close to a hospital. Furthermore we have at our disposal cheap pieces of technology such as mobile telephones, which make the communication between a home birth midwife and the local hospital team theoretically easy. In such a context the reasons to point out the differences between home birth and hospital birth tend to disappear. It is becoming easier to reconcile what the privacy of a home birth can offer and what the hospital facilities can offer.

In many cases the decision regarding the place of birth might be postponed until the onset of labour. If the progress of labour is straightforward, why not stay at home? If it seems wise to do so, finally give birth in hospital. By then one of the most common reasons for a long and difficult delivery – going to the hospital too soon, before established labour, at a phase where women are very sensitive to the effects of the environment – has been eliminated. Even women who do not feel comfortable with the concept of home birth should at least try to go to the hospital as late as possible, after a sort of point of no return. This is not easy in the context of the nuclear family, when the only adult around is the father, who is often dominated by the fear that the baby might be born before arrival at hospital. That is why, during the period of transition, in certain countries, doulas might play a key role.

DOULAS IN A PERIOD OF TRANSITION

A typical doula is a mother or grandmother who gave birth 'normally'. She is the mother figure a young woman can rely on during the whole period surrounding birth. The doula phenomenon appears as an aspect of the rediscovery of authentic midwifery. From the 1970s onwards this Greek word was used by John Kennell and Marshall Klaus in their studies of the presence of a lay female companion during labour. The Greek community does not like this term because the doula was a slave in ancient Greece. A midwife from Athens told me that she would prefer a term such as 'paramana', which means 'with the mother'. However we'll refer to doulas, because the term has been used in several published studies and is now well accepted.

John Kennell and Marshall Klaus started their studies in the 1970s in two busy hospitals in Guatemala, where 50 to 60 babies were born every day and where the routines had been established by doctors and nurses from the United States. They found that the presence of a doula dramatically reduces the incidence of all sorts of intervention and the use of drugs, and improves the outcome. They reproduced their studies in Houston, Texas, in a neighourhood where the population is predominantly Hispanic and incomes are low. The birthing care-givers there were directed by English-speaking residents in a 12-bed ward. The doulas spoke both Spanish and English. As in Guatemala, the presence of a doula had positive effects here, too.

As long as the studies were conducted in low-income Hispanic populations, the statistical results clearly confirmed the positive effects of the presence of a doula. The findings were different in the context of the Kaiser Permanente Care Program of Western California, where the presence of a doula had no impact on the rates of caesarean deliveries and other operative deliveries. Such differences need to be interpreted. At Kaiser Permanente, the population was typically middle-class American. In such a context the baby's father was almost always present. Unfortunately the authors of the report did not provide any information about the way the doulas had been chosen. They found it more important to underline that all of them attended approved training programmes and had served as doulas for at least two births under the supervision of a more experienced doula. One can wonder if the training cannot be counter-productive. Once I had dinner with three of the doulas who were involved in the Houston study. They spoke a lot about the birth of their own children as positive experiences. They never mentioned any training. The term 'training' suggests that what the doula does is more important than who she is. This does not mean that doulas should not be informed.

With a well-informed doula, the young mother feels more secure. An ideal doula must be aware of anything related to pregnancy, childbirth and breastfeeding, even if her know-ledge is superficial. Let us imagine a pregnant woman who heard a doctor mentioning the possibility of a placenta praevia: the doula must at least understand what it means.

An information session for doulas should focus on first aid in obstetrics, so that real emergencies, that are exceptionally rare, can be immediately recognised. For example if, after a sudden rupture of membranes, the cord is visible at the vulva, the doula should conclude that the need is to reach a hospital without any delay; on the way, she will contact the medical team and use the phrase 'prolapse of the cord'. If a pregnant woman has a sudden terrible abdominal pain, without any remission, and is in a state of shock, the doula will use the phrase 'suspicion of placenta abruptio' while contacting the medical team. If a baby is born in an unexpected place, for example in a car, the doula will know that in the case of such a quick and easy birth there is usually nothing to do. The only important point is to make sure that the place is warm enough and that mother and baby are not cold. Cutting the cord is not a physiological necessity. However, it is such a well-established ritual that many first aid programmes teach how to cut the cord.

The media often report the stories of heroic and clever dads who used their shoe laces and a pair of kitchen scissors in order to cut the cord. When a woman has a baby in such places as a train or a plane, the sensational report by the media usually focuses on the persons who were there by chance and who 'delivered' the baby, rather than on the woman who gave birth.

The future of the doula phenomenon depends upon how the word doula is understood. If the doula is still another person introduced in the birthing place in addition to the midwife, the doctor and the father, her presence will be counter-productive. If the focus is on the training of the doula rather than on her way of being and her personality, the doula phenomenon will be a missed opportunity.

NEW NATIVITY

During the period of transition when we are waiting for a new generation of low-profile midwives, some marginal women find a way to meet their need for absolute privacy during labour. They do not call any qualified person. They just give birth at home by themselves. Such women are not readily receptive to the usual depowering vocabulary suggesting that it is impossible to give birth without a helper, or a guide, or a coach, or a support person, or a partner, or any sort of trained health professional. They intuitively know that self-confidence plus absolute privacy create the best possible situation for an easy birth.

Although their behaviour is usually deemed incomprehensible and irresponsible, we can learn from these women. We must realise that in spite of millennia of culturally controlled childbirth, there are still women who are in touch with their most archaic mammalian needs. A biodynamic attitude to childbirth will be based on the deep-rooted needs of labouring women, not on the role of the birth attendants. Such women offer food for thought for those who will practice obstetrics or midwifery before 2032.

17 BEING A MIDWIFE OR AN

OBSTETRICIAN BEFORE 2032

UNADAPTABLE MIDWIVES

A certain number of midwives, all over the world, find it difficult to adapt to the current high degree of industrialisation of childbirth. They realise that, in conventional midwifery schools, they have been trained to follow protocols established by medical committees. They feel that they are the prisoners of a system and that this system is destroying the art of midwifery. Some of them just stop working, as if waiting for a new phase in the history of childbirth. Others play the Trojan horse and try to fight the system from inside. Others look for alternatives to the practice in conventional obstetric units. Some have expressed their need to be 'un-trained'.

During the phase of transition we are expecting, the concept of 'un-training midwives' might become less esoteric and less mysterious. The term has been introduced by Jeannine Parvati Baker, the mother of six and the founder of Hygieia College, based in Utah. The vision shared by Jeannine and her disciples is that every mother is a midwife. The international 'Conferences' organised by this 'Mystery School' are termed 'Gatherings' and usually held in the wilderness. Hygieia now has about 1000 students on the five continents. Let us hope that the objectives of the college will be better understood in the near future.

STILL IMPRISONED IN A SYSTEM

There are also obstetricians who feel that they are the prisoners of a system and who are trying to move into another kind of practice. I met some of them in a great diversity of countries. Some are really isolated. Others can satisfy their need to belong to a network of colleagues. I spent a day in Seoul with obstetricians who founded the group 'Better Birth'. They came either from Seoul or from divers remote parts of the rural areas of the country. Let us underline that in Korea the midwives have almost completely disappeared, women usually give birth in a highly medicalised environment, and the rates of caesareans are often around 40 per cent. In a country where they beat the world records for the number of ultrasound scans per pregnancy, it is significant that one of our main topics of conversation at dinner time was antenatal care. It appeared that antenatal care is an aspect of obstetrical practice that can be to a certain extent changed almost overnight, at least more rapidly than the attitudes and the environment during the birth itself.

The concept of antenatal 'care' is new. It appeared and developed during the twentieth century. It is therefore inseparable from the concepts of industrialised and medically controlled childbirth. This explains why the dominant style of antenatal care is constantly focusing on potential problems. Since the only objective of conventional prenatal visits is to detect abnormalities and pathological conditions, the word 'care' is inappropriate and even deceptive. Standardisation and routine tests are the main characteristics of industrialised antenatal care. Such characteristics can be illustrated by countless extracts from the medical literature. I offer the typical example of a short text published in the medical journal the *Lancet*. The author belongs to an Australian women and children's health institute. She wants to introduce a couple's visit in the framework of antenatal care. She regrets that 'men's attendance at antenatal clinics might take some years to become "routine" after explaining the reasons for involving the man: "Couples" visits should not be promoted as an HIV-related visit but as an opportunity to screen and treat for other infectious diseases, including sexually transmitted infections and tuberculosis, and to discuss emergency transport plans if complications arise during pregnancy or labour'. Obviously the point is to talk about risks – nothing else – as many times as possible and to as many people as possible.

In such a context, it's no wonder that antenatal visits often have a powerful 'nocebo effect', which is a negative effect on the emotional state of pregnant women and indirectly of their families. We must keep in mind that in many countries most women have about ten antenatal visits; in other words they have ten opportunities to hear about potential problems. Modern pregnant women cannot be blissful. All of them have at least one reason to be worried: 'your blood pressure is too high or too low', 'your weight is increasing too quickly or too slowly', 'you are anaemic', 'you might hemorrhage because your platelet count is low', 'you have a gestational diabetes', 'your baby is too small or too big', 'the placenta is low', 'you are 18 and teenage pregnancy is associated with specific risks', 'you are

39 and pregnancy at an old age is associated with specific risks', 'your baby has not yet turned head first', 'according to the blood sample you are at risk of having a Down's syndrome baby', 'you did not take folic acid at the right time and we must consider the risk of spina bifida', 'you are not immunised against rubella', 'you are Rh negative', 'you should have given birth last Wednesday, therefore we must consider an induction of labour', and so on. Is it still possible to be a 'normal' woman?

It is also on the subject of antenatal care that today the deep-rooted tendency of obstetrics to standardise, the key word 'routine', and therefore the conventional obstetrical attitudes in general are seriously challenged by the results of a series of authoritative epidemiological studies.

The RADIUS (Routine Antenatal Diagnostic Imaging with Ultrasound) trial involved more than 15,000 pregnant women. It compared those who, at random, had routine ultrasound screening with those who had only ultrasound scans on demand (when it was needed to resolve a particular problem). The number of scans per woman per pregnancy in the group with on-demand ultrasonography was finally 0.6.

The conclusion, as expressed in an editorial of the *New England Journal of Medicine*, is clear: there is no medical reason to offer routine ultrasound scans to the overall population of pregnant women.

There are similarities between this study and a huge study, on the entire Canadian population, revealing that the routine use of the tests to detect gestational diabetes does not improve the outcomes. After the publication of this study, the American College of Obstetrics and Gynecology updated its guidelines, claiming that it is not malpractice not to routinely screen for gestational diabetes. Let us add the conclusions of a trial by WHO involving 53 centres in Thailand, Cuba, Saudi Arabia and Argentina. This trial confirmed that effective antenatal care can be provided with fewer visits than are presently made routinely.

A quick review of these recent studies explains why a certain number of obstetricians are currently torn between two opposing influences. On the one hand they have been trained at the height of industrialised childbirth. They are still the prisoners of a system based on the watchwords 'routine' and 'protocol'. On the other hand their ingrained common sense is aroused by an accumulation of scientific data that leads them to reconsider the standardisation of antenatal care. Are we on the way towards a real biodynamic attitude?

BIODYNAMIC ATTITUDE AND ANTENATAL CARE

A real biodynamic attitude would imply a radically different style of prenatal care. In 1991 and 1992, in the framework of research about nutrition during pregnancy in a large conventional London hospital, I realised that even in such an environment it would be easy to change the dominant attitude in prenatal care. I interviewed 500 pregnant women, selected at random.

The interview was an addition to the standard antenatal consultation. The main part of our conversation was about the specific nutritional needs of the developing brain of the baby. Most of these women were eager to talk about the growth of their baby and had a tendency to lengthen the interview. This anecdote suggests that health professionals can have a positive effect on the emotional state of pregnant women. One of their main preoccupations would be at least to have a protective effect.

In the current scientific context we are in a position to understand how certain maternal emotional states – particularly those associated with high levels of hormones such as cortisol – can influence the development of the foetus. Furthermore dozens of studies, included in our data bank (www.birthworks.org/primalhealth), confirm that the emotional state of a pregnant woman has lifelong consequences on her child. It would not take so long for an adaptable health professional to shift towards a positive attitude.

Although it should be mostly the role of the family and the community, certain health professionals may encourage and even organise activities that have a direct positive effect on the emotional state of pregnant women. In a French hospital, in the 1970s and 1980s, we bought a piano so that, on Tuesday evenings, pregnant women could come and sing together. Everybody was invited to participate in these popular evenings, including the fathers, the midwives, the cleaning ladies, the secretary, etc. At the end of these singing sessions it was obvious that everybody was happy. We could assert – without measuring the levels of cortisol and catecholamines! – that the hormonal balance of the pregnant women was favourable for the growth and development of their baby in the womb.

In fact the first duty of the health practitioner should be at least to reduce the risk of having a 'nocebo effect' while communicating with a pregnant woman. It is almost always possible – and justified – to present the results of a clinical exam or a test in a positive way.

Let us take the example of a pregnant woman who has been told, at the end of her pregnancy, that she has an increased blood pressure. The current disease-oriented attitude probably led to present this result as bad news. The message has been that there is something wrong. A biodynamic attitude, on the other hand, would lead to an explanation that a simple increased blood pressure is an adaptive response; it should not be confused with pre-eclampsia, which is a disease associating a high blood pressure with protein in the urine and various metabolic disturbances. An isolated increased blood pressure is usually a good sign of placental activity. The placenta – as the advocate of the baby – manipulates maternal physiology via the release of hormones and asks the mother to provide more blood.

There are at least four authoritative studies demonstrating that a 'pregnancy induced hypertension' is associated with good outcomes. Practitioners who want first to protect the emotional state of a pregnant woman know how to use reassuring analogies in order to explain that a sign, such as high blood pressure, should not be confused with a disease, such as pre-eclampsia. For example, 'when you have a brain tumour, you have a headache; but when you have a headache, it does not mean that you have a brain tumour'. It can be still more reassuring to tell the pregnant woman that there is no need to measure her blood pressure again, since ultimately only the presence of protein in the urine is the prerequisite for the diagnosis of pre-eclampsia.

A practitioner whose priority is to protect the emotional state of pregnant women would never use a scaring term with a strong nocebo effect such as 'gestational diabetes'. If a modification of the metabolism of sugars has been detected, he (she) will explain that it is a transitory response: the baby has sent a message to the mother, via the placenta, in order to receive more sugar. The mother's body has adapted to this demand by becoming less sensitive to the effects of insulin. Gestational diabetes has been called 'a diagnosis still looking for a disease'. When the diagnosis has been established, it leads to simple recommendations that should be given to all pregnant women, such as: avoid pure sugar (soft drinks, etc.); prefer complex carbohydrates (pasta, bread, rice, etc.); have a sufficient amount of physical exercise. In fact a biodynamic attitude would lead to testing the glucose tolerance in only a very small number of selected cases.

I constantly receive phone calls from pregnant women who are in a state of anxiety – even of panic – after an antenatal visit. One of the most common reasons for such phone calls is a misinterpretation of the concentration of haemoglobin (the pigment of the red blood cells) which is routinely measured. Many practitioners think that iron deficiency in pregnancy can be detected via the haemoglobin concentration. When a woman has a result in the region of 9.0 or 9.5 at the end of her pregnancy, more often than not she is told that she is anaemic and she is given iron tablets. She understands that there is something wrong in her body that needs to be corrected.

If, on the other hand, the practitioner is anxious to protect her emotional state, is interested in placental physiology and reads the medical literature, she will be given good news. She will be told that, according to huge statistics, results in the region of 9.0 are associated with the best possible birth outcomes. It will be explained to her that the blood volume of a pregnant woman is supposed to increase dramatically, and that the haemoglobin concentration indicates the degree of blood dilution. She will understand that the results of her tests are suggestive of effective placental activity and that her body is responding correctly to the instructions it is given. The placenta asks the mother to make her blood more fluid. Once more there is a tendency to confuse a transitory physiological response (blood dilution) and a disease (anaemia).

A widespread misunderstanding of placental physiology is at the root of such misinterpretations. The lack of interest in the role of the placenta as an advocate of the baby is as deep-rooted as the lack of interest in birth physiology.

Physiologists study the universal laws of nature. Scorning the physiological perspective is a characteristic of industrialised childbirth ... and industrialised farming as well. Understanding the laws of nature and working with them, on the other hand, is the main characteristic of a biodynamic attitude, whether in plant cultivation, animal breeding or childbirth. The evolution of antenatal care gave us an opportunity to illustrate and clarify the concept of a biodynamic attitude.

18   GETTING OUT OF THE CUL-DE-SAC

Today the need to radically transform agriculture, animal breeding and husbandry is widely understood. Solutions are available. It is also theoretically feasible to minimise the use of fossil energies in order to control, before it is too late, the massive emission of greenhouse gases and its climatic consequences.

The varieties of Homo that are currently occupying the planet are so clever that they can formulate the most complex problems and conceive solutions. But the same varieties of *Homo sapiens* are also characterised by an inability to recognise the real priorities and to put certain ideas into action. That is why humanity is in a cul-de-sac.

10,000 YEARS AGO

Homo sapiens entered this cul-de-sac about 10,000 years ago, with the advent of agriculture and with the domestication of animals. The 'neolithic revolution' spread out from several avant-garde societies in places such as the Middle East, South East Asia, Central China, Central America and the Andes. From that time the basic strategy for survival of all human groups has been to dominate nature. The domination of nature – which implies the concept of property – became a major cause of conflicts. War became a common aspect of the relationship between human groups. Humans groups developed a tendency to dominate and even to eliminate each other. From that time it became an advantage to develop the human potential for aggression.

This turning point in the history of humanity is also a time when our ancestors were given opportunities to improve their understanding of human reproduction. The observation of domesticated animals made clear the role of sexual intercourse and therefore the role of the male. Genital sexuality became organised and controlled via different matrimonial arrangements and via a great variety of rituals, including genital mutilations. Since that time all the different episodes of sexual life have been under the control of the cultural milieu. This is the case with childbirth.

The transmission of beliefs is a powerful way to control the birth process and particularly the phase of labour between the birth of the baby and the delivery of the placenta. Let us refer again to the cross-cultural belief that colostrum is tainted or harmful – even a substance to be expressed and discarded. This negative attitude towards colostrum implies that, immediately after being born, the baby must be in the arms of another person than the mother. This is the origin of a widespread deep-rooted ritual, which is to rush to cut the cord.

We cannot make a comprehensive list of all known rituals that meddle in the newborn baby's relationship with his or her mother. We cannot either mention all the beliefs that reinforce the common attitude towards colostrum. This is the case, for example, of the belief shared by several West African ethnic groups that the mother should not look at the newborn's eyes, so that the bad spirits cannot enter the baby's body. What are the evolutionary advantages of this multitude of beliefs and rituals that tend to challenge the maternal protective instinct during a short period of time considered critical in the development of the capacity to love?

In the current scientific context, we think of asking the questions that way, because answers can be suggested. From the time when the basic strategy for survival of most human groups was to dominate nature and to dominate other human groups, it was an advantage to make human beings more aggressive and able to destroy life. In other words it was an advantage to moderate the capacity to love, including love of nature, that is to say the respect for Mother Earth.

It is understandable that since the beginning of the neolithic revolution the most successful societies were those who had at their disposal the most appropriate beliefs and rituals in the period surrounding birth. Our interpretation is confirmed by data from a very small number of pre-agricultural peoples that could be studied before becoming extinct, and who had other strategies for survival. Their strategy was to live in perfect harmony with the ecosystem.

The priority in these societies was not to develop the human potential for aggression. We know in particular about childbirth among the !Kung San, a pre-agricultural African human group. When birth seemed imminent, the woman used to walk a few hundred yards, to find an area in the shade, to clear it, to arrange a soft bed of leaves, and to give birth on her own.

Over the millennia there has been a selection of human groups according to their potential for aggression. We are all the fruits of such a selection. This explains our inability to recognise and to take action against clear-cut manifestations of impaired capacity to love. That is why we cannot easily get out of the cul-de-sac.

TODAY

Today the ultimate priority is not to transform farming or to moderate the emission of greenhouse gases. It is to make possible the advent of another variety of *Homo*. This variety of *Homo* – the authentic *Homo sapiens* – must be able to invent new strategies for survival at a time when the limits to the domination of nature become obvious. He (she) must be able to wonder how the respect for Mother Earth develops. He (she) must be able to participate in a dialogue between humanity and Mother Earth, which implies a certain degree of unification of humanity. In other words he (she) must master the energies of love.

If the planet is to sustain human life in the future, we must prepare for a non-genetic mutation initiated by necessity, reason and scientific knowledge. Such a mutation is not utopian in the age of the scientification of love. We are learning that the capacity to love develops through a long chain of early experiences, particularly in the period surrounding birth. The way babies are born is the critical link of the chain that is routinely disturbed. It is also the link of the chain on which it is possible to act. After some millennia of culturally controlled childbirth, the current variety of *Homo* is not yet damaged enough to miss the possible mutation towards the authentic *sapiens*.

That is why the current industrialisation of childbirth should become the main preoccupation of those interested in the future of humanity. Let us dream that the vital objective of Hygieia College will be shared by millions of human beings within decades: 'Healing the Earth by Healing Birth'.

FURTHER READING

CHAPTER 1
THE LAST STRAW

Bader, F., Davis, G., Dinowitz, M., Garfinkle, B., Harvey, J., Kozak, R., Lubiniecki, A., Rubino, M., Schubert, D., Wiebe, M. and Woollett, G. (1998) Assessment of risk of bovine spongiform encephalopathy in pharmaceutical products, *Biopharm.* January, pp. 20–31.

Bradley R. (1999) BSE transmission studies with particular reference to blood, *Dev. Biol. Stand.*, 99:35–40.

Brown, D.W.G. (2001) Foot and mouth disease in human beings, *Lancet* 357:1463.

Brown, P., Cervenakova, L., McShane, L.M., Barber, P., Rubenstein, R. and Drohan, W.N. (1999) Further studies of blood infectivity in an experimental model of transmissible spongiform encephalopathy, with an explanation of why blood components do not transmit Creutzfeldt-Jakob disease in humans, *Transfusion* 39:1169–78.

Brown, P., Rohwer, R.G., Dunstan, B.C., MacAuley, C., Gajdusek, D.C. and Drohan, W.N. (1998) The distribution of infectivity in blood components and plasma derivatives in experimental models of transmissible spongiform encephalopathy, *Transfusion* 38:810–16.

Donnelly, C.A., Ghani, A.C., Ferguson, N.M. and Anderson, R.M. (1997) Recent trends in the BSE epidemic, *Nature* 389:903.

FDA (2000) TSE Advisory Committee, transcript of June meeting.

Harris, D.A. (1999) Cellular biology of prion diseases, *Clin. Micro. Rev*, 12:429–44.

Holden, Patrick (2000) Howard's way, *Living Earth* 211:14.

Kimberlin, R.H. (1991) An overview of bovine spongiform encephalopathy, *Dev. Biol. Stand.* 75:75–82.

Prempeh, H., Smith, R. and Muller, B. (2001) Foot and mouth disease: the human consequences, *BMJ* 322:5656.

Taylor, D.M., Fraser, H., McConnell, I., Brown, D.A., Brown, K.L., Lamza, K.A. and Smith, G.R.A. (1994) Decontamination studies with the agents of bovine spongiform encephalopathy and scrapie, *Arch. Virol.* 139:313–26.

USDA (United States Department of Agriculture), consult <www.usda.gov>

Venters, G.A. (2001) New variant Creutzfeldt-Jakob disease: the epidemic that never was, *BMJ* 323:858–61.

Wilesmith, J.W., Wells, G.A.H., Ryan, J.B.M., Gavier-Widen, D. and Simmons M.M. (1997) A cohort study to examine maternally-associated risk factors for bovine spongiform encephalopathy, *The Vet Record* 141:239–43.

CHAPTER 2
MAJOR PREOCCUPATIONS AND LATEST SPECTACULAR EVENTS

Alaluusua, S., Lukinmaa, P.-L., et al. (1993) Exposure to 2,3,7,8-tetrachlorodibenzo-para-dioxin leads to defective dentin formation and pulpal perforation in rat incisor tooth, *Toxicology* 8:1–13.

Alaluusua, S., Lukinmaa, P.-L., et al. (1996) Polychlorinated dibenzo-p-dioxins and dibenzofurans via mother's milk may cause developmental defects in the child's teeth, *Environ. Toxicol. Pharmacol.* 1:193–7.

Alaluusua, S., Lukinmaa, P.-L., et al. (1999) Developing teeth as biomarker of dioxin exposure, *Lancet* (16 January) 353:206 (research letter).

Allan, B.B., Brant, R., Seidel, J.E. and Jarrel, J.F. (1997) Declining sex ratios in Canada, *Can. Med. Assoc. J.* 156:37–41.

Astolfi, P. and Zonta, L.A. (1999) *Hum. Reprod.* 14(12):3116–19.

Auger, J., Kunstmann, J.M., Czyglik, F. and Jouannet, P. (1995) Decline in semen quality among fertile men in Paris during the past 20 years, *N. Engl. J. Med.* 332:281–5.

Brown, C., Heitkamp, M., et al. (1981) Niacin reduces paraquat toxicity in rats., *Science* 212:1510–11.

Chkraborty, D., Bhattacharya, A., et al. (1978) Biochemical studies of polychlorinated biphenyl toxicity in rats: manipulation by vitamin C. *Int. J. Vitamin Nutr. Res.* 48:22–31.

Clark, D. and Prouty, R. (1997) Experimental feeding of DDE and PCB to female big brown bats, *J. Toxicol. Environ. Health* 2:917–28.

Davis, D.L., Gottlieb, M.B. and Stampnitzky, J.R. (1998) Reduced ratio of male to female births in several industrial countries. A sentinel health indicator? *JAMA* 279:1018–23.

deFreitas, A. and Norstrom, R. (1974) Turnover and metabolism of polychlorinated biphenyls in relation to their chemical structure and the movement of lipids in pigeons, *Can. J. Physiol.* 52:1080–94.

Dimich-Ward, H., Hertzman, C., et al. (1998) Reproductive effects of paternal exposure to chlorophenate wood preservatives in the sawmill industry, *Scand. J. Work Environ. Health* 24(5):416.

Feitosa, M.F., Krieger, H. (1992) Demography of the human sex ratio in some Latin American countries, 1967–1986, *Hum. Biol.* 64:523–30.

Forman, D. and Moller, H. (1994) Testicular cancer, *Cancer Surv.* 19–20:323–41.

Garcia-Rodriguez, J., Garcia-Martin, M., et al. (1996) Exposure to pesticides and cryptorchidism: geographical evidence of a possible association, *Environ. Health Perspect.* 104:394–9.

Hassold, T. (1983) Sex ratio in spontaneous abortions, *Ann. Hum. Genet.* 47:39–47.

Huisman, M., Koopman-Esseboom, C., et al. (1995) Neurological condition in 18-month-old children perinatally exposed to polychlorinated biphenyls and dioxins, *Early Human Development* 43:165–76.

Infante-Rivard, C. and Sinnett, D. (1999) Preconceptional paternal exposure to pesticides and increased risk of childhood leukaemia, *Lancet* 354:1819 (letter).

Jackson, M.B. (1988) John Radcliffe Hospital cryptorchidism research group: the epidemiology of cryptorchidism, *Horm. Res.* 30:153–6.

Jacobson, J.L. and Jacobson, S.W. (1996) Intellectual impairment in children exposed to polychlorinated biphenyls in utero, *N. Engl. J. Med.* 335(11):783–9.

Jacobson, J.L. and Jacobson, S.W. (2001) Postnatal exposure to PCBs and childhood development, *Lancet* 358:1568–9.

Krstevska-Konstantinova, M., Charlier, C., Craen, M., Du Caju, M., Heinrichs, C., de Beaufort, C., Plomteux, G. and Bourguignon, J.P. (2001) Sexual precocity after immigration from developing countries to Belgium: evidence of previous exposure to organochlorine pesticides, *Hum. Reprod.* May 16(5):1020–6.

Lambert, G. and Brodeur, J. (1976) Influence of starvation and hepatic microsomal enzyme induction of the mobilization of DDT residues in rats, *Tox. App. Pharm.* 36:111.

Marcus, M., Kiely, J., McGeehin, M. and Sinks, T. (1998) Changing sex ratio in the United States, 1969–1995, *Fertil. Steril.* 70(2):270–3.

Mizuno, R. (2000) The male/female ratio of fetal deaths and births in Japan, *Lancet* 356:738–9.

Mocarelli, P., Brambilla, P., et al. (1996) Change in sex ratio with exposure to dioxin, *Lancet* 348:409.

Mocarelli, P., Gerthoux, P.M., et al. (2000) Paternal concentrations of dioxin and sex ratio of offspring, *Lancet* 355:1858–63.

Moller, H. (1996) Change in male–female ratio among newborn infants in Denmark, *Lancet* 348:828–9.

Nelson, B.K., Moorman, W.L. and Shrader, S.M. (1996) Review of experimental male-mediated behavioral and neurochemical disorders, *Neurotoxicol. Teratol.* 18(6): 611–16.

Olshan, A.F. and Faustman, E.M. (1993) Male mediated developmental toxicity, *Annual Rev. Public Health* 14:159–81.

Paulozzi, L.J., Erickson, D. and Jackson, R.J. (1997) Hypospadias trends in two US surveillance systems, *Pediatrics* 100:831.

Primal Health Research (1999a) newsletter, Spring, 6(4).

Primal Health Research (1999b) newsletter, Autumn, 7(2).

Sever, L.E. (1995) Male mediated developmental toxicity, *Epidemiology* 6:573–4.

Sharpe, R.M. and Skakkebaek, N.E. (1993) Are estrogens involved in falling sperm counts and disorders of the male reproductive tract? *Lancet* 341:1392–5.

Stowe, C. and Plaa, G. (1958) Extrarenal excretion of drugs and chemicals, *Am. Rev. Pharmacol.* 8:337–56.

Tielemans, E., Van Kooij, R., et al. (1999) Pesticide exposure and decreased fertilisation rates in vitro, *Lancet* 354:484–5.

van der Pal-de Bruin, K.M. (1997) Change in male–female ratio among newborn babies in the Netherlands, *Lancet* 349:62.

Vartiainen, T., Kartovaara, L. and Tuomisto, J. (1999) Environmental chemicals and changes in sex ratio: analysis over 250 years in Finland, *Environ. Health Perspect.* 107:813–15.

Walkowiak, J., Wiener, J.A., Fastenbend, A., et al. (2001) Environmental exposure to polychlorinated biphenyls and quality of the home environment: effects on psychodevelopment in early childhood, *Lancet* 358:1602–7.

Wiemmels, J.L. and Gazzaniga, G., et al. (1999) Prenatal origin of acute lymphoblastic leukemia in children, *Lancet* 354:1499–503.

Wirth, A., Schlierf, G. and Schelter, G. (1979) Physical activity and lipid metabolism, *Klin. Wochenschr.* 57:1105–201.

CHAPTER 3
THE SOURCE AND THE TARGET

Blanchette, I. and Dunbar, K. (1999) Memory for analogies and analogical inferences, in *Proceedings of the Twenty First Annual Meeting of the Cognitive Science Society*, pp. 73–7.

Blanchette, I. and Dunbar, K. (2000) How analogies are generated: the roles of structural and superficial similarity, *Memory & Cognition* 28:108–24.

Dunbar, K. (1997) How scientists think: online creativity and conceptual change in science, in Ward, T.B., Smith, S.M. and Vaid, S. (eds) *Conceptual Structures and Processes: Emergence, Discovery and Change*, APA Press, Washington DC (also published in Japanese in 1999).

Dunbar, K. (1999) The scientist invivo: how scientists think and reason in the laboratory, in Magnani, L., Nersessian, N. and Thagard, P. *Model-based Reasoning in Scientific Discovery*, Plenum Press, pp. 89–98.

Dunbar, K. and Baker, L.M. (1994) Goals, analogy, and the social constraints of scientific discovery, *Brain & Behavioral Sciences* 17:538–9.

Gentner, D., Holyoak, K.J. and Kokinov, B. (eds) (2000) *Analogy: Perspectives from Cognitive Science*, MIT Press.

Holst, Glendon (2001) There is no Aha! in a brute force search (A CPSC 449 Honours Thesis) Department of Computer Science, University of British Columbia.
Consult <www.cs.ubc.ca/spider/gholst/HonoursThesis/.html>

Schunn, K. and Dunbar, K. (1996) Priming, analogy and awareness in complex reasoning, *Memory and Cognition* 24:271–84.

Usha, Goswami (1998) *Cognition in Children*, Psychology Press.

CHAPTER 4
SIMILARITIES

Carter, Jenny and Duriez, Thereze (1986) *With Child. Birth through the ages*, Mainstream.

Donnison, Jean (1977) *Midwives and Medical Men*, Heinemann.

McNetting, Robert (1993) *Smallholders, Householders*, Stanford University Press.

Odent, Michel (1992) *The Nature of Birth and Breastfeeding*, Bergin and Garvey.

Wertz, Richard and Wertz, Dorothy (1989) *Lying-in: A History of Childbirth in America*, Yale University Press.

CHAPTER 5
ENTHUSIASM

Abramsky, L., Botting, B., Chapple, J. and Stone, D. (1999) Has advice on periconceptional folate supplementation reduced neural-tube defects? *Lancet* 354:998.

Jordan, Miriam (2001) Routine surgery: for Brazilian women, cesarean sections are surprisingly popular, *Wall Street Journal* 14 June.

Leavitt, Judith Walzer (1986) *Brought to Bed: Childbearing in America*, Oxford University Press.

Mathers, C.D., Sadana, R., Salomon, J.A., Murray, C.J.L. and Lopez, A.D. (2001) World Health Report 2000: Healthy life expectancy in 191 countries, *Lancet* 357: 1685–91.

CHAPTER 6
REMEMBER THEM!

Conford, Philip (2001) *The Origins of the Organic Movement*, Floris Books.

Gaskin, Ina May (2001) *Spiritual Midwifery* 4th edition, The Farm Book Publishing Company.

Leboyer, Frederick (1995) *Birth Without Violence*, Healing Art Press (reprint).

McCarrison, Robert (1961) *Nutrition and Health*, Faber and Faber.

Reich, Wilheim (1953) *The Murder of Christ*, Farrar, Straus and Giroux.

Reich, Wilheim (1983) *Children of the Future*, Farrar, Straus and Giroux (first published in the *Orgone Energy Bulletin*).

Steiner, Rudolf. Information in English about Rudolf Steiner and anthroposophy can be obtained from the Anthroposophical Society of Great Britain, Rudolf Steiner House, 35 Park Road, London NW1 6XT.

Wrench, G.T. (1972) *The Wheel of Health*. Shocken Books.

CHAPTER 8
WHICH DISASTER ARE WE WAITING FOR?

Bookchin, Murray (1974) *Towards an Ecological Society*, Black Rose Books.

Cnattingius, S., Hultman, C.M., Dahl, M. and Sparen, P. (1999)

Very preterm birth, birth trauma, and the risk of anorexia nervosa among girls, *Arch. Gen. Psychiatry* 56:634–89.

Forssman, H. and Thuwe, I. (1981) Continued follow-up study of 120 persons born after refusal of application for therapeutic abortion, *Acra. Psychiatr. Scand.* 64:142–9.

Hattori, R., et al. (1991) Autistic and developmental disorders after general anaesthesic delivery, *Lancet* 1 June, 337:1357–8 (letter).

Huttunen, M. and Niskanen, P. (1978) Prenatal loss of father and psychiatric disorders, *Arch. Gen. Psychiatr.* 35:429–31.

Jacobson, B. and Bygdeman, M. (1998) Obstetric care and proneness of offspring to suicide as adults: case control study, *BMJ* 317:1346–9.

Jacobson, B. and Nyberg, K. (1990) Opiate addiction in adult offspring through possible imprinting after obstetric treatment, *BMJ* 301:1067–70.

Jacobson, B., Nyberg, K., et al. (1987) Perinatal origin of adult self destructive behaviour, *Acra. Psychiatr. Scand.* 76:364–71.

Kubicka, L., Matejcek, Z., et al. (1995) Children from unwanted pregnancies in Prague, Czech Republic, revisited at age thirty, *Acra. Psychiatr. Scand.* 91:361–9.

Lynskey, M., Degenhardt, L., Hall, W. (2000) Cohort trends in youth suicide in Australia 1964–1997, *Aust. NZ J. Psychiatry* June, 34(3):408–12.

Myhram, A., Rantakallio, P., et al. (1996) Unwantedness of a pregnancy and schizophrenia of a child, *Br. J. Psychiatr.* 169:637–40.

Nyberg, K., Buka, S.L. and Lipsitt, L.P. (2000) Perinatal medication as a potential risk factor for adult drug abuse in a North American cohort, *Epidemiology* 11(6):715–16.

Odent, M. (1986) *Primal Health*, Century Hutchinson.

Odent, M. (2000) Between circular and cul-de-sac epidemiology, *Lancet* 355:1371.

Raine, A., Brennan, P. and Medink, S.A. (1994) Birth complications combined with early maternal rejection at age 1 year predispose to violent crime at 18 years, *Arch. Gen. Psychiatry* 51:984–8.

Salk, L. and Lipsitt, L.P., et al. (1985) Relationship of maternal and perinatal conditions to eventual adolescent suicide, *Lancet* 16 March, pp. 624–7.

Tinbergen, N. and Tinbergen, A. (1983) *Autistic Children*, Allen and Unwin.

CHAPTER 9
THE SCIENTIFICATION OF LOVE

Odent, M. (2001) *The Scientification of Love* 2nd edition, Free Association Books.

Pedersen, C.A. and Prange, J.R. (1979) Induction of maternal behavior in virgin rats after intracerebroventricular administration of oxytocin, *Pro. Natl Acad. Sci.* 76:6661–5.

CHAPTER 10
BEES

Amyes, Sebastian G.B. (2002) *Magic Bullets, Lost Horizons: The Rise and Fall of Antibiotics*, Horwood Academic Press.

Darwin, Charles (undated) *Fertilisation of Orchids by Insects*, C. M. Coleman.

Delaplane, K.S., and Mayer, D.F. (2000) *Crop Pollination by Bees*, CABI Publishing.

Gorbach, S.L. (2001) Antimicrobial use in animal feed – time to stop, *N. Engl. J. Med.* 345:1202–3.

McDonald, L.C., Rossiter, S., et al. (2001) Quinupristin-Dalfopristin resistant enterococcus faecium on chicken and in human stool specimens, *N. Engl. J. Med.* 345:1155–60.

White, D.G., Zhao, S., et al. (2001) The isolation of antibiotic resistant salmonella from retail ground meats, *N. Engl. J. Med.* 345:1147–54.

CHAPTER 11
FALLING ASLEEP AND FALLING IN LABOUR

Odent, M. (1990) Position in delivery, *Lancet* 12 May, 335(8698):1166.

Odent, M. (1996) Knitting needles, cameras and electronic fetal monitors, *Midwifery Today.* Spring, 37:14–15.

Odent, M. (1997) Preparing for the post-electronic birthing age, *Midwifery Today* Fall, 43:19–20.

Odent, M. (1998a) Active versus expectant management of third stage of labour, *Lancet* 30 May, 351(9116):1659.

Odent, M. (1998b) Don't manage the third stage of labour! *Pract. Midwife* 1 September, 9:31–3.

Odent, M. (2000) Insights into pushing: the second stage as a disruption of the fetus ejection reflex, *Midwifery Today* Fall, 55:12.

Odent, M. (2001) New reasons and new ways to study birth physiology, *Int. J. Gynecol. Obstet.* November, 75 Suppl 1:S39–S45.

CHAPTER 12
IS THE PARTICIPATION OF THE FATHER AT BIRTH DANGEROUS?

Odent, M. (1986) The world of men and the world of women, in *Primal Health*, Century Hutchinson.

CHAPTER 16
HAVING A BABY BEFORE 2032

Gordon, N.P., Walton, D., et al. (1999) Effects of providing hospital-based doulas in health maintenance organization hospitals, *Obstet. Gynecol.* 93(3):422–60.
Kennell, J., Klaus, M., et al. (1991) Continuous emotional support during labor in a US hospital, *JAMA* 265:2197–201.

CHAPTER 17
BEING A MIDWIFE OR AN OBSTETRICIAN
BEFORE 2032

Berkowitz, R.L. (1993) Should every pregnant woman undergo ultrasonography? *N. Engl. J. Med.* 329:874–5.
Bucher, H.C. and Schmidt, J.G. (1993) Does routine ultrasound scanning improve outcome in pregnancy? Meta-analysis of various outcome measures, *BMJ* 307:13–17.
Curtis, S., et al. (1995) Pregnancy effects of non-proteinuric gestational hypertension, SPO Abstracts, *Am. J. Obst. Gynecol.* 418:376.
Ewigman, B.G., Crane, J.P., et al. (1993) Effect of prenatal ultrasound screening on perinatal outcome, *N. Engl. J. Med.* 329:821–7.
Garn, S.M., et al. (1981) Maternal hematologic levels and pregnancy outcome, *Semin. Perinatol.* 5:155–62.
Hygieia. Consult <www.freestone.org/hygieia.html>
Jarrett, R.J. (1993) Gestational diabetes: a non-entity? *BMJ* 306:37–8.
Jarrett, R.J., Castro-Soares, J., Dornhorst, A. and Beard, R. (1997) Should we screen for gestational diabetes? *BMJ* 315:736–9.

Kilpatrick, S. (1995) Unlike pre-eclampsia, gestational hypertension is not associated with increased neonatal and maternal morbidity except abruption, SPO abstracts, *Am. J. Obstet. Gynecol.* 419:376.

Koller, O., Sandvei, R. and Sagen, N. (1980) High hemoglobin levels during pregnancy and fetal risk, *Int. J. Gynaecol. Obstet.* 18:53–6.

Naeye, E.M. (1981) Maternal blood pressure and fetal growth, *Am. J. Obstet. Gynecol.* 141:780–7.

Steer, P., Alam, M.A., Wadsworth, J. and Welch, A. (1995) Relation between maternal haemoglobin concentration and birth weight in different ethnic groups, *BMJ* 310:489–91.

Symonds, E.M. (1980) Aetiology of pre-eclampsia: a review, *J. R. Soc. Med.* 73:871–5.

Von Dadelszen, P. and Ornstein, M.P. et al. (2000) Fall in mean arterial pressure and fetal growth restriction in pregnancy hypertension: a meta-analysis, *Lancet* 355:87–92.

Villar, J. and Ba'aqueel, H., et al. (2001) WHO antenatal care randomized trial for the evaluation of a new model of routine antenatal care, *Lancet* 357:1551–64.

INDEX

Compiled by Sue Carlton